Seriation, Stratigraphy, and Index Fossils

The Backbone of Archaeological Dating

Seriation, Stratigraphy, and Index Fossils

The Backbone of Archaeological Dating

Michael J. O'Brien

and

R. Lee Lyman

University of Missouri–Columbia
Columbia, Missouri

Kluwer Academic / Plenum Publishers
New York, Boston, Dordrecht, London, Moscow

ISBN: 0-306-46152-8 (Hardbound)

233 Spring Street, New York, N.Y. 10013

10 9 8 7 6 5 4 3 2 1

A C.I.P. record for this book is available from the Library of Congress

Printed in the United States of America

Preface

It is difficult for today's students of archaeology to imagine an era when chronometric dating methods—radiocarbon and thermoluminescence, for example—were unavailable. How, they might ask, were archaeologists working in the preradiocarbon era able to keep track of time; that is, how were they able to place objects and sites in proper sequence and to assess the ages of sites and objects? Given the important roles that chronometric methods play in modern archaeology, it is little wonder that today's students might view earlier efforts to establish chronological ordering as imprecise and unworthy of in-depth study. This is unfortunate, because even casual perusal of the large body of literature that resulted from the efforts of archaeologists working during the first half of the twentieth century reveals that they devised a battery of clever methods to determine the ages of archaeological phenomena, often with considerable precision.

This kind of chronological control is referred to colloquially as *relative dating*: production of a sequence of events for which no fixed or calendric dates exist. Instead of knowing that a certain kind of pottery was made between, say AD 100 and AD 300, and that another kind was made between AD 300 and AD 600, all we know is that the latter kind is of more recent origin than the former. The latter kind could postdate the earlier kind by several hundred years or by a thousand years, but we do not know this; all we know is that it is more recent. In like manner, we might know, perhaps through historical evidence, the terminal calendric date of manufacture and use of the later kind of pottery, but we might not know when on a calendric scale that kind was first made, and thus when it began replacing the earlier kind.

Numerous methods for working out relative chronological orderings have been devised in archaeology, one of which, stratigraphic excavation, had its roots in geological observations of the eighteenth century. Stratigraphic excavation is perhaps the best known of the various relative-dating methods used by prehistorians, no doubt because the majority of the archaeological record has a geological mode of occurrence. There are as well two other methods—seriation and cross dating—that deserve special consideration. These three general methods each comprise numerous techniques that prehistorians earlier this century worked out in order to keep track of time. Usually these are treated in cursory fashion in general archaeology texts, and the history of their development is summarized in superficial, and often inaccurate, fashion. Thus our goals in this volume are two. We seek first to describe in some detail the various ways each method works. Second, in order to understand how the various techniques for implementing each method were developed and why they work the way they do, we place our discussion within the historical context of its development, focusing particularly on what happened in North America. Our reason for doing so is simply that we believe a detailed understanding of the history of a method or technique can clear up various misunderstandings and omit ambiguities, both in terms of analytical technique and disciplinary history.

There are several techniques of using artifacts from superposed strata to measure time, but these are rarely, if ever, differentiated; the more common practice is to lump them under the heading "stratigraphic excavation." As we will see, there is considerable disparity in terms of exactly what stratigraphic excavation means. Culture historians of the early twentieth century often are credited with bringing about a "stratigraphic revolution," generally taken to mean that it was not until a few years after 1910 that prehistorians began to excavate stratigraphically and to make observations concerning superposition. However, the *real* revolution was in *how* prehistorians generated information from the items gathered from the excavation of superposed strata, not the fact that they were excavating stratigraphically. That had been routine procedure for decades. But without a firm understanding of the history behind relative-dating methods and of the myriad changes that the methods went through over time, it is easy not only to conflate various methods and techniques but also to muddy that history of who did what, when they did it, why they did it, and how it differed from what came before it.

Our personal interest in dating methods grew out of our larger and more general interest in the culture history period of Americanist archaeology, which extended from about 1910 to 1960. We produced three volumes on that period: a book that examines the archaeology of a particular state in terms of how it reflected general trends in Americanist archaeology (O'Brien, 1996a), a reader containing what we consider benchmark papers (Lyman, *et al.*, 1997a), and a detailed account of why culture history took the form that it did (Lyman *et al.*, 1997b). Our interest in the culture history paradigm in turn grew out of the appreciation that American prehistorians working in the first half of the twentieth century went through many of the same contortions and engaged in many of the same kinds of epistemological discussions that contemporary evolutionary biologists do. Always in the background of archaeological discussion was the notion of culture development and how prehistorians could best structure their work to understand culture change as it was manifest in the archaeological record. Not surprisingly then, time came to have a lead role in archaeologists' efforts to understand the past, since any interpretation of cultural development had to be constructed around that variable. Until about 1915, temporal ordering of archaeological phenomena was rarely accomplished because of the general belief that there was minimal temporal depth to the archaeological record of the Americas, and what little there was comprised cultures that were not very different from those documented by early explorers, settlers, and anthropologists.

We mark the middle of the second decade of the twentieth century as the turning point in Americanist archaeology in terms of when it became scientific. It was at that point that archaeology could be wrong and *know* it was wrong relative to matters of chronology. Archaeologists wanted to be scientific in how they studied cultural phenomena, and they began borrowing heavily, but implicitly, from other fields, especially evolutionary biology. Many of the archaeological schemes that grew up during the culture historical period, for instance, taxonomic schemes for classifying archaeological phenomena, were loosely constructed analogues borrowed wholesale from biology with little or no thought given to their applicability to archaeological data, or lack thereof.

As archaeologists interested in evolution, we view our roots as grounded heavily in culture history. The problems with which culture historians wrestled, such as how to order phenomena chronologically, as well as how to sort homologies from analogies, were as germane to archaeology as they were, and still are, to paleontology and evolutionary biology. In this volume we do not discuss the deep connections between modern evolutionary archaeology and culture history [we do that elsewhere in considerable detail (Lyman and O'Brien, 1997; O'Brien and Lyman, n.d.)] but rather we examine not only the various chronological methods that arose within the intellectual climate of culture history but also some of the assumptions that underlay their formation and development.

We have attempted to produce a book that will appeal to all generations of archaeologists, from graduate students to seasoned professionals. Our desire to bridge the gap in expertise necessitated careful consideration of what to include and what to leave out. The danger in such an exercise is that one group will perhaps want to see more historical discussion, whereas another will be more interested in how relative-dating methods are used today, if indeed they are. Many of the case examples that we use to highlight the methods come from the culture historical period. It is easy to find modern examples of stratigraphy, since stratigraphic observation is still at the center of what archaeologists do, and is relatively easy to do for cross correlation given that many archaeological types have become index fossils, but it is difficult to find modern examples of seriation. Several modern studies make it sound as if this method still plays a significant role in Americanist archaeology, only because the term *seriation* now includes a variety of techniques and methods that are but remotely related to what seriation was early in the twentieth century.

We gratefully acknowledge the advice and assistance of our editor at Plenum, Eliot Werner. All figures were drafted by Dan Glover, who tracked down various source materials, prepared the references, and made numerous other significant contributions to the project. E. J. O'Brien read the manuscript in its entirety and edited it for content as well as for style; he also wrote the discussion of cosmological time in Chapter 1. This is the fourth of our books on which he has worked, and we greatly appreciate the help and advice he has provided over the years. Finally, we acknowledge an intellectual debt of gratitude to Bob Dunnell for his advice and constructive criticism over the last two decades.

Contents

Chapter 1. An Introduction to Time and Dating 1

Preliminary Considerations 5
Relative and Absolute Time 8
Continuous and Discontinuous Time 9
Direct and Indirect Dating 11
Scientific Dating 12
Time and Its Measurement 13
Nominal Scale Measurement 17
Ordinal Scale Measurement 17
Interval Scale Measurement 18
Ratio Scale Measurement 19
Beyond Measurement Scale: Ideational and Empirical Units 21

Chapter 2. The Creation of Archaeological Types 23

Initial Considerations 24
Constructing Chronological Types 26
Chronological Types in Americanist Archaeology 32
Typological Issues Begin to Take Shape 33
Typology in Retrospect 57

Chapter 3. Seriation I: Historical Continuity, Heritable Continuity, and Phyletic Seriation 59

What Is Seriation? 60
Seriation in Americanist Archaeology 62
The Key Assumptions: Historical and Heritable Continuity 65
Continuity and the Study of Organisms 67
Tracing Lineages 72
Detecting Heritable Continuity 74
Historical Continuity, Heritable Continuity, and the Study of Artifacts 80
Culture Traits 82
W. M. Flinders Petrie and Artifacts from Egyptian Tombs 84

John Evans and Gold Coins from Britain 91
A. V. Kidder and Pottery from Pecos Pueblo 94
The Gladwin–Colton–Hargrave System 96
Projectile Point Evolution 101

Chapter 4. Seriation II: Frequency Seriation and Occurrence Seriation 109

The First Frequency Seriation 111
How Do Occurrence and Frequency Seriation Work? 114
The Seriation Model 116
Requirements and Conditions of Seriation 117
Occurrence Seriation 119
Frequency Seriation 121
Meeting the Conditions of the Seriation Model 125
Temporal Resolution and Rates of Change 130
Absolute Seriation 132
A Final Note 136

Chapter 5. Superposition and Stratigraphy:
Measuring Time Discontinuously 139

Strata, Stratigraphy, and Superposition 144
Stratigraphic Excavation 147
Stratigraphic Excavation in Historical Context 149
Early Stratigraphic Excavation 151
On the Eve of the "Revolution" 157
The Real Revolution 158
What *Was* the Revolution? 171
After the Revolution: Measuring Time with Strata 172
Measuring Time at Gatecliff Shelter, Nevada 175
The Final Proof Is in the Spade, But 180

Chapter 6. Cross Dating: The Use of Index Fossils 185

Folsom and Clovis Points 188
George C. Vaillant and the Mexican Formative 191
James A. Ford and the Lower Mississippi Valley 199
Measuring Time Discontinuously 212

Chapter 7. Final Thoughts on Archaeological Time:
A Clash of Two Metaphysics ... 217

Measuring Time Continuously ... 219
Measuring Time Discontinuously ... 221
Concluding Remarks ... 225

References ... 227

Index ... 247

Seriation, Stratigraphy, and Index Fossils

The Backbone of Archaeological Dating

1 An Introduction to Time and Dating

Archaeologists traditionally have been interested in three aspects of the archaeological record: Where things come from (space), what they look like (form), and how old they are (time). Without denigrating the study of either space—the "where"—or form—the "what"—we would venture to say that placing archaeological specimens, commonly referred to as *artifacts*, in their proper chronological position is probably the most fundamental exercise in archaeology. As anthropologist Berthold Laufer (1913:577) put it, chronology "is at the root of the matter, being the nerve electrifying the dead body of history." Maya prehistorian Alfred Tozzer (1926:283) put it only slightly differently: "[A]rchaeological data have an inert quality, a certain spinelessness when unaccompanied by a more or less definite chronological background." Without the ability to produce a temporal order, archaeology is reduced to a study of synchronic variation in which objects are examined simply for their functional, stylistic, and/or aesthetic characteristics. All these aspects of artifacts are important components of archaeological inquiry, but at its core archaeology is the study of change. Without methodologically sound ways to measure the passage of time, the archaeological record is reduced to a jumble of materials that might as well all date to a single point in time because we cannot study change without a way to arrange those materials in time.

As important a role as chronological ordering plays in modern archaeology relative to the study of change, this role is a comparatively recent phenomenon in Americanist archaeology, where it dates primarily to the period after about 1915. What was afoot during earlier periods that kept the study of time so firmly in the background? Part of the answer resides in the nonscientific nature of Americanist archaeology in the nineteenth century—a nature far removed from that which characterized contemporary European archaeology, especially that conducted in England. With certain key exceptions (e.g., Thomas, 1894), the majority of studies conducted by Americanist prehistorians of the nineteenth century could best be described as antiquarianism. Archaeological sites were explored, often in a casual manner, and in some cases artifacts were described in resulting reports, though rarely in anything other than cursory fashion. Artifacts were assigned to various known cultural groups that had been documented ethnographically, but there was little interest in placing the materials in any kind of chronological order.

During the latter half of the nineteenth century, not all Americanist archaeology was antiquarianism, but oddly enough it was the scientifically minded prehistorians more than the antiquarians who retarded the growth of archaeological interest in time. The major reason for the disinterest in time was the notion that little time had elapsed since North America was first inhabited by people. To be sure, there was a vocal minority opinion—that humans had been in North America since at least the end of the last major glacial age—but that view was consistently held in check by prehistorians connected with the leading anthropological institution of the period, the Bureau of American Ethnology (BAE).

The BAE was created by Congress in 1879, and its first directive was to identify the group or groups responsible for constructing the thousands of large mounds and enclosures scattered over the eastern United States. Had they been constructed by ancestors of the American Indians or was an extinct race of mound builders responsible, a group that later had been driven out of North America by the Indians? John Wesley Powell, the first director of the BAE, hired Cyrus Thomas to solve the problem, which he and members of his Division of Mound Exploration eventually did (Thomas, 1884, 1891, 1894) by demonstrating a cultural connection between the mound builders and contemporary Native American groups. Powell's strategy was to use archaeology to extend the history of the American Indian back in time, and to him and his colleagues at the BAE, that history was extremely short, perhaps on the order of 1000 years or so. Importantly, that history did not include a separate race of mound builders, nor did it have room for "the idea of an earlier, unrelated Paleolithic 'race' " (Meltzer and Dunnell, 1992:xxxiii).

Prehistorians not connected with the BAE did not share this atemporal view, and it was they who argued that the gravel beds of the midwestern and eastern United States contained rich *in situ* deposits of tools similar to Paleolithic artifacts from Europe that occurred in what everyone knew to be glacial-age contexts. Since the implements looked similar, so the thinking went, they must be similar in age. The literature of the late nineteenth century is replete with articles that come down on one side or the other of the issue. The leading opponent of the notion of a human presence in North America during the end of the last glacial period was BAE prehistorian William Henry Holmes, who carefully and critically undermined the position of proponents by demonstrating that equivalency in artifact form did not necessarily indicate equivalency in age, that "gravel deposits" containing artifacts were not necessarily of glacial age, and that artifacts within glacial-age deposits could be intrusive to those deposits (e.g., Holmes, 1892, 1893, 1897). Proponents of the glacial-age propositions were a mixture of respected geologists and prehistorians, the latter including Frederic Ward Putnam of the Peabody Museum (Harvard), Warren K. Moorehead of the Phillips Academy in Andover, Massachusetts, and Thomas Wilson of the US National Museum.

How could a group of experts in the fields of geology and archaeology reach such different conclusions over the age of the first inhabitants of North America?

In a series of critical examinations of the intellectual history of the late nineteenth and early twentieth centuries, David J. Meltzer (1983, 1985, 1991) considered the differences in outlooks brought to bear on the question, concluding that they were natural outgrowths of educational differences:

> Most BAE archaeologists had earned their scientific credentials on geological and natural history expeditions to the deep canyons and vast Plains of the American West. This education was a wholly democratic American experience: anyone could be a scientist given a sufficient amount of energy and intellect. Formal training in a European university was not only unnecessary, it was viewed with a certain amount of disdain. The result was an innovative and highly nationalistic archaeology, with the American Indian as its centerpiece....
>
> This was not the case among archaeologists outside the BAE, particularly those in the universities, which closely followed European lines. Contemporary European archaeologists viewed their discipline as an extension of history into prehistoric times (Trigger, 1978). The strategy was historical rather than anthropological, with attention to sequence and detail. [Meltzer, 1985:251 (see also Meltzer, 1991)]

The so-called European strategy would eventually make its way into Americanist archaeology during the late nineteenth century, but it would not be until the second decade of the twentieth century that "sequence and detail" became part and parcel of archaeological method.

Meltzer's observations are important to understanding how Americanist archaeology started as it did, especially with regard to the study of time. Bureau personnel were not part of an intellectual history grounded in the investigation of deep, stratified sequences of prehistoric remains. To be sure, they were not ignorant of what was going on archaeologically in Europe, nor were they unmindful of geological principles. They were, however, conditioned to looking at the ethnographic present and interpreting the past in terms of it. Thus the cultural past was seen as nothing more than a quite recent ancestor of the present and was not significantly different from the cultural present documented by early explorers, settlers, and anthropologists. In contrast, contemporary European archaeologists viewed the past not only as an extension of the present and recent past into prehistoric times but as a deep extension. They quickly adopted the historical methods of geologists and paleontologists to understand the details of that sequence (Grayson, 1983; Van Riper, 1993).

We label the compressed view of time held by BAE personnel the *flat-past perspective*, though a better term would be the *shallow-past perspective*, since they recognized that at least a little time had passed since humans first arrived on the continent. Many historians of Americanist archaeology (e.g., Browman and Givens, 1996; Rohn, 1973; Strong, 1952; Willey, 1968; Willey and Sabloff, 1993) indicate that a "stratigraphic revolution" during the second decade of the twentieth century was the catalyst that prompted the emergence of an interest in time, but as we point out throughout this book, this view is incorrect. These historians overlook the fact that many archaeologists prior to 1910 were excavating in a manner that is readily considered stratigraphic. The problem was that the artifact assemblages they were excavating were described in terms that rendered the detection of chronological differences among them invisible, and thus archaeologists rarely asked chronological questions (Lyman and O'Brien, 1999). For Americanist archaeology to emerge as a scientific line of inquiry, one with empirically testable implications, archaeologists had to change the way they were measuring time. That is, they had to change how they categorized artifacts in order to answer their chronological questions. They accomplished this change in the second decade of the twentieth century, but that important switch was lost because historians focused on a change in excavation strategies that supposedly took place.

Prior to 1915, most Americanists were searching for differences in sets of culture traits of the sort that would suggest major qualitative differences in cultures. Culture traits were tallied in technological and/or functional terms—pottery, fishing gear, weapons of war or the hunt, and so forth—and thus cultures could be construed as occupying different temporal positions in a manner that aligned with a progressive evolutionary model of cultural development such as that proposed by Lewis Henry Morgan (1877). Cultural differences were therefore like those being reported in Europe; anything of less magnitude was insignificant, though certainly not improbable (e.g., Kroeber, 1909). This value system would change abruptly between 1914 and 1916 when archaeologists modified their scale of observation from that of the presence–absence of cultural traits to that of the frequencies of trait variants, or what would become known as artifact "styles."

Part of the reason for the change in value systems was that the little time afforded a human presence in North America was beginning to expand as prehistorians became persuaded that the archaeological record was too extensive and varied to be the result of a short human presence there. For example, Nels C. Nelson (1916) observed that there simply were too many ruins in the Southwest. If they had been constructed all at once, then incredibly large populations had lived in the region simultaneously—populations much larger than had been documented ethnographically in the last quarter of the nineteenth century. Thus, there *must* be more time depth to the record than was suspected previously. As more field work was carried out in parts of the East and Southwest, the dam of resistance against a deep antiquity of the American archaeological record sprang new leaks, finally rupturing in 1927, when, in a small *arroyo* in northeastern New Mexico, stone spear points similar to those in Fig. 1.1 were found among the remains of clearly ancient bison (Figgins, 1927). With that discovery, the prehistory of the continent took on a totally different complexion. As A. V. Kidder (1936b:143) noted, the discipline now faced "a paradox, for while the upper end of our time scale is being compressed [by dendrochronology and other techniques for measuring time on a calendric scale], its lower end is being vastly expanded by the recent unequivocal determination of high antiquity of Folsom Man."

The race was on to fill in the culturally unknown period between the end of

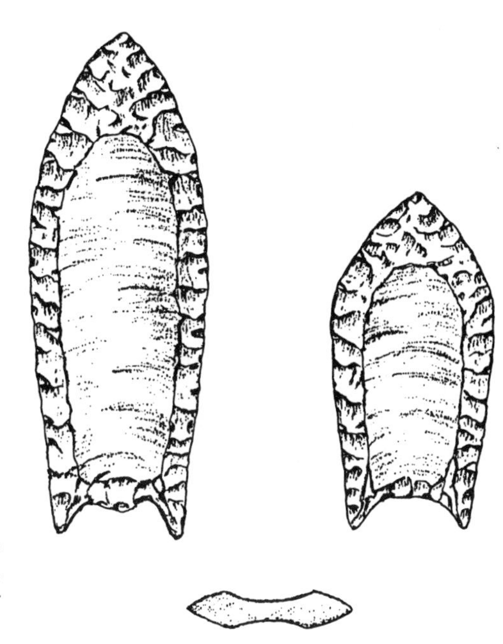

Figure 1.1. Folsom points from the Lindenmeier site, Larimer County, Colorado. Specimen at left is 6.8 cm long (after Roberts, 1935).

the Pleistocene (about 11,000 years ago) and the relatively recent past. Archaeologists knew they had two ends of the continental sequence: one marked by spear points and extinct animals, the other by such things as standing architecture, pottery, and any number of other traits captured in an ethnographer's notebook. But how did one go about putting all the stuff in between in chronological order? And how did one know that the chronological ordering into which things were placed was the *correct* order? The methods devised to ensure correct chronological order and that form the topic of this book were used to fill the gap and were developed a decade before the bottom dropped out of the shallow past.

PRELIMINARY CONSIDERATIONS

Throughout this book we use the term *dating* to refer to "the placement in time of events relative to one another or to any established scale of temporal measurement" (Dean, 1978:225). It has been suggested that "a *date* is a specific point in time, whereas an *age* is an interval of time measured back from the present," and that the former term, "when used as a noun, carries a connotation of calendar years and a degree of accuracy that is seldom appropriate" (Colman *et al.*, 1987:315). The latter term refers in part to the "chronological sensitivity" (Dean,

1993:60), or temporal resolution, of the dating technique used. We use the terms *date* and *age* in a manner similar to that suggested by Colman *et al.* (1987), though we do not always find this possible given the chronological sensitivity of the dating methods we discuss. An *event* can comprise the manufacture of an arrowhead, the use of a ceramic vessel, the deposition of a bone awl, or the caching of a tool kit (we provide a more detailed consideration of the concept of *event* in Chapter 2). Knowing the age or date of archaeological events, we can determine to various degrees of resolution the succession of events, the duration of an event, the simultaneity of events, or some combination thereof. We can also, if we so desire, determine the rate, or tempo, of change over particular spans of time and perhaps even detect the kind, or mode, of change.

We focus on three kinds of chronological ordering methods out of the repertoire that archaeologists have at their disposal. We picked these three—seriation, superposition and stratigraphy, and cross-dating through the use of index fossils—for several reasons, not the least of which is that together they have long formed the backbone (Tozzer's "spine") of archaeological dating. They are essential to almost any kind of archaeological analysis and should be familiar to any student who enters the discipline. We bypass some relative dating methods, including several geoarchaeological ones such as beach and terrace dating, varve dating, dating by rates of sediment accumulation, and dating by rates of chemical change in sediments. The methods we discuss are those that we find to be the most universally applicable; thus, they are the most useful ones, but they are also, as it turns out, the most misunderstood.

By concentrating on relative dating methods we bypass an entire suite of methods that in more modern times—the post-1950 era—have become popular in archaeology. These include radiocarbon and luminescence dating, two of the most widely applied methods, but also methods with more exotic-sounding names such as amino acid racemization, fission track dating, potassium–argon dating, and electron spin resonance. A plethora of books on these dating methods have been published over the past three decades. Some of these cover multiple dating methods (e.g., Aitken, 1990; Fleming, 1976; Geyh and Schleicher, 1990; Michael and Ralph, 1971; Michels, 1973; Roth and Poty, 1989; Taylor and Aitken, 1997), others cover a single chronometric or absolute dating method (e.g., for radiocarbon, see Berger and Suess, 1979; Bowman, 1990; Lowe, 1997; Taylor, 1987; Taylor *et al.*, 1992; for dendrochronology, see Baillie, 1982; for thermoluminescence, see Aitken, 1985; for archaeomagnetism, see Eighmy and Sternberg, 1990; for fission track dating, see Wagner and van den Haute, 1992). What we find surprising is that only one of these books on dating (Michels, 1973) devotes any space to relative dating, and that is a minimal 63 of 218 pages of text. During the preparation of this volume, we failed to find a book devoted to one or several of the relative dating methods available to archaeologists. The best we did was to find articles scattered here and there. We refer to these throughout the volume, and encourage readers to examine those articles for additional details.

The volume is organized into separate chapters that deal with each of the three dating methods, but the division is more out of convenience than anything else. For example, we cover the various techniques of the seriation method in two chapters because of the complexity of the method and its diverse means of implementation. Under each method we discuss its history (particularly in the Americas), how it works, and so on, but the problem with such an arrangement is that it might appear that each method is based on its own set of principles, and thus is self-contained. Such is not the case. None of the chronological methods discussed here was created in a vacuum, and all played off one another throughout the development of Americanist archaeology. It is fair to say that each method played a major role in the creation of a distinctly Americanist archaeology that was in several ways quite different from contemporary European archaeology (Lyman *et al.*, 1997b, 1998; Lyman and O'Brien, 1999). For example, although some textbooks treat seriation as a European invention that was borrowed by Americanists working in the Southwest, frequency seriation—the particular kind of seriation that most introductory textbooks treat—was an American invention.

The coalescence of seriation, superposition and stratigraphy, and cross-dating into an integrated body of methods devised to track the enormous variation in the North American archaeological record resulted in the appearance of a new paradigm for Americanist archaeology—culture history—which, despite considerable rancor over method and technique, was a fairly coherent entity from about 1915 to about 1960 (Lyman *et al.*, 1997a,b). It was this paradigm with which many archaeologists of the 1960s became fed up, so much so that they argued for the development of a new paradigm, one based on the search for cultural processes that underlay the creation of the archaeological record. Yet, the fundamental principles of the culture history paradigm were so ingrained in Americanist thought that archaeologists, whether they realized it or not, carried many of them over to the new paradigm (Meltzer, 1979). Those principles remain so ingrained in our thinking today that we often fail to acknowledge them or to realize that we use them day in and day out as we go about our research.

Although the literature might lead one to suspect otherwise, Americanist archaeology's attempts to become scientific did not originate with the processual archaeology of the 1960s and 1970s. Early twentieth-century Americanists expressed a similar desire, and how they attempted to fulfill that desire is an interesting part of our discipline's history (Lyman *et al.*, 1997a,b). Importantly, because those attempts still influence Americanist archaeology decades later, it is reasonable to ask whether earlier culture historians attained their goal, and, if not, how and why they failed. Always in the middle of culture historians' efforts to make archaeology a science was the notion of culture change: how best to detect, measure, and explain it. And at the center of any discussion of change is the notion of time and how to measure it.

Relative and Absolute Time

What we treat here often are referred to as *relative dating methods*, meaning that they measure time but with no indication of the amount of time that elapsed between each pair of events and no indication of when on a calendric scale the events occurred. Relative dating methods provide only a chronological sequence of events. In contrast, *absolute dating methods*, sometimes referred to as *chronometric* methods, yield the amount of time that elapsed between each pair of events as well as a calendric date indicating when each event occurred and perhaps each event's duration as well. Absolute dating methods thus provide much more than a simple chronological sequence. Such methods are preferred because they produce more detailed, higher resolution, chronological information and allow rates of change and durations of periods to be determined. This is because the duration of a temporal unit such as a solar year, also termed the sidereal, or astronomical, year, and the typical calendric unit employed is the same regardless of where it occurs along the temporal continuum.

Although the term *absolute* is a generally accepted modifier for the word *time*, we should ask, absolute in terms of what? No dating technique is, strictly speaking, "absolute," given how dates are produced (Colman *et al.*, 1987:315). Time is relative because to be perceived it must be *related* to a scale. Absolute dates "are expressed as points on standard scales of time measurement" (Dean, 1978:226), whereas relative dates are expressed in relation to one another. As Albert C. Spaulding (1960:447) observed some years ago,

> Relative time scaling is simply ranking an event before or after some other event. Absolute time scaling means placing an event with respect to a sequence of events which are thought to occur at regular intervals and which are given a standard designation by reference to an arbitrarily chosen initial point. Our absolute is, of course, the calendar.

Therefore, *all* time is relative, whether to the origin of the universe or to some other reference point. A distinction different from relative versus absolute time is quantifiable [Colman *et al.* (1987:315) suggest "numerical"] versus nonquantifiable [Spaulding (1960:448) suggests "time ranking"] time. Although we follow traditional archaeological practice and use the terms relative and absolute throughout the book, keep the quantifiable–nonquantifiable dichotomy in the back of your mind. What we mean here is well captured by the concept of scales of measurement (Stevens, 1946), a subject we return to in more detail later in this chapter. Here, a brief introduction will suffice.

Nominal scales of measurement record differences in kind, such as the scale of male and female, or the scale of citizenship such as Italian, Australian, and so on.

Ordinal scales of measurement are those that record differences of greater than and less than, but not the magnitude of difference. The geologist's hardness scale, known as "Mohs hardness scale," for minerals records that diamond is harder (Mohs scale = 10) than orthoclase (Mohs scale = 6), which in turn is harder than gypsum (Mohs scale = 2). But the difference in hardness between diamonds and orthoclase is not the same as the difference between orthoclase and gypsum, despite the fact that the differences between the two pairs of symbolic values on the scale—10 and 6 and 6 and 2—are mathematically identical. *Interval scales* measure not only greater than or less than relation, but also the magnitude of the difference, and thus are often thought of as *quantitative* (Stevens, 1946:679). Temperature is a well-known interval scale measure; the difference between 40°C and 39°C is exactly the same as the difference between 32°C and 31°C.

Relative dating methods produce relative dates, or a sequence of events placed on an ordinal scale. Event A can be said to be older than event B, but we cannot say how much older A is than B; that is, we cannot say when on a calendric scale A and B occurred. Absolute dating methods produce absolute, chronometric, or calendric dates, or sequences of events aligned on a scale of units in which each unit is of the same duration as every other unit. If event A is determined to be 10 years old and event B is determined to be 20 years old, then we can say that event B is 10 years older than event A, and we can also say that B is twice as old as A. Of course, we can only say such things if each of the units, in this case solar years, on the scale are of precisely equivalent duration.[1] It should be clear, then, that it is critically important to keep relative (ordinal scale) time and dating methods distinct from absolute (interval scale) time and dating methods during analysis of the archaeological record. Archaeologists have not always kept the two separate, resulting in such things as the "Radiocarbon Revolution" (Renfrew, 1973; see also Taylor, 1996), when the absolute dates assigned to various events dated by relative dating methods were found to be far off the mark.

Continuous and Discontinuous Time

Another distinction that should be kept in mind is of that between viewing time as continuous and viewing it as discontinuous. Both views have played important roles in archaeology, though the distinction between them has rarely been noted. Conflation of the two has caused severe problems that continue to plague archaeological analysis. But, you might think, a clock does not stop; thus time can hardly be discontinuous. Time, you say, is like the flight of an arrow

[1]As Dean (1978:226) points out, a radiocarbon year "is not a fixed entity"; that is, some radiocarbon years are of longer durations than a solar year, some are shorter than a solar year, and some are of durations equivalent to a solar year. Dean's point is that the calendric scale of radiocarbon years comprises units (radiocarbon years) of varying durations, and thus radiocarbon dating produces neither relative nor absolute dates, but something in between—a mixture of ordinal and interval scales.

(Gould, 1987): unidirectional and never stopping (ignoring, for the sake of the metaphor, the force of gravity on the arrow's flight). What, then, is meant by the terms *continuous* time and *discontinuous* time? In short, we are distinguishing between two different methods and results of measuring time. Time is measured as a sequence of unique phenomena that mark moments in time, whether those phenomena are the shifting ratio of decayed to nondecayed atoms of an isotope, the shrinkage of fins on an automobile or fish, or the altered positions of planets and stars. Time, in other words, is measured relative to change in, or the alteration of, phenomena.

Change, by definition, takes place over time. If there is no change, then this does not mean that time does not pass or that the clock of time stops. What it does mean, however, is that the phenomenological clock—the clock of change—against which the clock of time is arrayed stops. The potential result, then, is a failure to measure change, and thus time is rendered discontinuous, because phenomenological stasis results from how we perceive and record the alteration of phenomena against which time's clock is set. In archaeology, time typically is arrayed against or measured relative to the phenomenological clock of culture change. Certainly a culture may not change, that is, it may be static in structure and composition, over some duration of time. Potential problems arise when the apparent stasis of a culture is equated with a portion of the temporal continuum, a portion we might designate as a unit termed a *period* (Rowe, 1962b). Depending on how the boundaries of such temporal units are identified phenomenologically, change within particular periods can be masked, rendering culture change or the phenomenological clock discontinuous relative to time's flow, or the clock of time. In such cases all change takes place at the instant in time that serves as a boundary between periods (Plog, 1974). Time's flow in the sense in which archaeologists are interested in it and in terms of how they measure it—culture change—is rendered discontinuous. Throughout the book we touch on the problems that result from dating methods that measure time discontinuously.

Before moving on, we offer an argument from a different context to warrant our suggestion that although time is a continuum it can be measured discontinuously. In his book *Time's Arrow, Time's Cycle*, paleobiologist Stephen Jay Gould (1987) reviews some of the history of how geologists came to measure and conceive of time (see also Kitts, 1966). His point is that time can be viewed as either equivalent to the linear flight of an arrow or as cyclical. How can the latter view emerge from the study of earth history? Simply by using a device for measuring time, or more correctly the alteration of phenomena, that is built to detect cycles. The biblical creation, destruction, and recreation of the earth is one example. The tectonic building of mountain ranges, their subsequent erosion, and their later reemergence from level plains as the plates of the earth's surface continue to collide is another. The clock of change never stops but cycles through a set of stages that geologists derived as empirical generalizations from their observations of geological processes: time's cycle. The clock of time ticks ever onward during the operation of these processes—time's arrow—but we perceive cycles, given the methods we use to measure time and mark its passage:

> Time's arrow of "just history" marks each moment of time with a distinctive brand. But we cannot, in our quest to understand history, be satisfied only with a mark to recognize each moment and a guide to order events in temporal sequence. Uniqueness is the essence of history, but we also crave some underlying generality, some principles of order transcending the distinction of moments.... We also need, in short, the immanence of time's cycle. (Gould, 1987:196)

Granting that time can appear cyclical if it is measured in particular ways, the next conceptual step—that time can appear to be discontinuous if it is measured in other ways—should not be a difficult one to take.

Direct and Indirect Dating

The dating methods now available to archaeologists are, as we near the end of the second millennium, numerous and varied in how they work. Some of them are used to date phenomena directly, whereas others are used to date phenomena indirectly; some can be used both ways depending on the chronological question being asked. Generally, archaeologists want to know when an artifact was made, used, or deposited, or when a house was built, occupied, or abandoned. When such comprise the events of interest, they are what Dean (1978:228) refers to as the "target event." The age of any one of these events can be determined by studying either the attributes of the event itself or the attributes of an event temporally associated with the event of interest. These two kinds of study are generally referred to as *direct dating* and *indirect dating*, respectively. Direct dating methods involve measurement of the attributes of the event of interest and assigning that event a relative or absolute age based on the values of the measured attributes. Indirect dating methods involve measurement of attributes displayed by an event that is associated with the target event; an age is assigned to the "dated event" (Dean, 1978:226) based on its attributes, and the age of the target event is inferred to be the same based on Worsaae's principle of association (Rowe, 1962c).

This principle of association, named after Danish prehistorian J. J. A. Worsaae (1849), holds that two or more things found in the same depositional unit, or stratum, are likely to be of similar age. If a gray pot with red triangles painted on it and a side-notched projectile point are found in the same stratum, they can be inferred to be of the same age; that is, they were made, used, and deposited at approximately the same time. If we know the pot was made sometime during the period AD 300–500, then we can infer that the projectile point was made during that same time period. Given what we today know about how the archaeological record has been formed (Schiffer, 1987), it should be clear that the inference of contemporaneity is just that—an inference. The projectile point could be intrusive to the stratum containing the pot. Detailed consideration of this and other such problems with indirect dating methods are largely beyond the scope of this book, but we touch on them where critical to the discussion (see Dean, 1978, for more details).

Scientific Dating

It was not only with the dawn of the atomic age that archaeology was able to date an object or phenomenon in an absolute sense. Archaeologists have long realized, for example, that objects of known manufacture date can provide chronological control. This was one method Cyrus Thomas used to demonstrate that some of the mounds in the Southeast were being used when Europeans arrived. Another absolute dating method, dendrochronology, or tree ring dating, has its roots (no pun intended) deep in Western science. Astronomer A. E. Douglass often is credited with founding dendrochronology as a solid scientific enterprise; as a result of his efforts in the American Southwest beginning in 1913, tree ring dating became a chronological method with archaeological utility (Douglass, 1929). This recognition, however, overlooks the fact that one of the first, if not *the* first, papers on the use of tree rings to date archaeological remains was published in 1838, by Englishman Charles Babbage. A few years later, Americans Ephraim Squier and Edwin Davis used tree rings to affix dates to some of the mounds they included in their "Ancient Monuments of the Mississippi Valley" (Squier and Davis, 1848).

The majority of absolute dating methods currently in use in archaeology postdate the beginning of the atomic age. Radiocarbon dating was developed in the late 1940s by Willard Libby of the University of Chicago and became available for general use in the early 1950s (Libby, 1955; Taylor, 1985). It revolutionized archaeology by providing absolute chronological anchor points for the relative sequences that archaeologists had long been creating. Although radiocarbon was a boon to the discipline, it soon created an unfortunate result: Archaeologists began relying on radiocarbon dates and forgetting the basics of the relative dating methods we discuss in this book.

As opposed to the radiocarbon method, which can work only on once-living things, luminescence dating works on inorganic materials. The principle behind luminescence dating has long been known, but it was only in the 1970s that the method, particularly thermoluminescence, began to be used widely (Aitken, 1985). Since it can be used to date inorganic objects such as pottery directly as opposed to dating organic remains recovered in association with inorganic artifacts, thermoluminescence offers considerable advantages over radiocarbon. One drawback is the high cost associated with obtaining dates, a cost that can exceed by several times that associated with obtaining routine radiocarbon dates.

Relative dating, on the other hand, is relatively inexpensive. There is no cost involved in making stratigraphic observations or in seriating collections of pottery sherds, other than the salary of the person(s) making the observations or seriating the collections. Are such dating methods less scientific because they are performed outside a laboratory full of the apparatus many of us associate with scientific analysis? No, and in fact seriation, stratigraphic observation, and cross-dating possess all the rigor that any machine-centered method does, as well as all the problems in interpretation. The decline in use that relative dating methods, especially seriation, have experienced in Americanist archaeology has in large part been the result of a confidence in methods that seem to be more scientific than the three discussed here. In our opinion much of this confidence is misplaced. Simply because an absolute date is derived by way of chemistry and physics does not make it more scientific than a relative date derived by way of seriation. If the hallmark of scientific research is that its results are testable, then relative dating is readily subsumed under the umbrella of science.

Archaeologists routinely argue that what they really want to know is not simply that projectile point A is older than point B but how *much* older. Relative dating, it is argued, is fine when one is taking an initial stab at constructing a chronology but not when one wants to know *when* something happened in the past. To know that, one needs machines to assign dates to objects and events. But before we relegate relative dating methods to the margins of archaeological inquiry, we had better understand them and be clear about some of the epistemological pitfalls that surround the use of all dating methods, regardless of whether they are relative or absolute. Further, as Bonnie Blackwell and Henry Schwarcz (1993:56) indicate, "we should not abandon relative dating methods in this era of absolute dating advances. [The former continue] to offer critical criteria against which any [absolute] date must be evaluated." In other words, as Luther Cressman (1951:311) observed early on, the radiocarbon dating "method is no miraculous tool for the archaeologist.... [H]e is still going to have to depend on sound stratigraphic [and other relative dating] methods" and reasoning. Radiometric assays may lend an air of authority to a dating exercise, but this matters little if the dates are interpreted incorrectly as a result of faulty logic. Dating methods, regardless of how scientific they might appear, can do nothing to correct errors in logic. In the remainder of this book, we want to examine closely how best to avoid such errors. The best place to begin is with time itself and the ways in which it can be measured.

TIME AND ITS MEASUREMENT

Archaeologist Albert Spaulding (1960:447) remarked that "Time itself is a continuum sensed as a succession of events." This statement says several very important things. Time is continuous, it is an abstraction, and it is perceived as a sequence of phenomena. The first phrase indicates that time "is not packaged, but rather infinitely divisible" (Ramenofsky, 1998:75). Consequently, in conjunction with the second phrase, "chronologies cannot be discovered" (Ramenofsky, 1998:75) but instead must be constructed. How? As Spaulding (1960:448) put it, "[a]ll chronological judgments are inferences made by interpreting spatial and formal attributes in the light of physical, biological, or cultural principles." That is, we must devise ways to measure and record variation in phenomena. The measurement devices must record that variation in units that themselves can be inferred to represent the passage of time; because the units are only inferentially chronological, their ability to measure the passage of time must be testable. These devices provide the means by which time is measured as variation in phenomena, and thus are the basis for the construction of a chronology of events: "Chronological units are conceptual, defined, and imposed on the continuum of time" (Ramenofsky, 1998:75). Time is a concept that is measured with tools we call chronologies (Ramenofsky, 1998:74), which in turn are lists of varying phenomena recorded and measured with chronological units. We reserve discussion of the units used to construct chronologies for Chapter 2.

If we had to sift through the various kinds of logical errors commonly made with respect to chronology and single out the most significant one, it probably would be the conflation of different measures of time. The sources of such confusion are easy to identify. For example, Colman *et al.* (1987:317) distinguish six kinds of dating methods on the basis of the assumptions and mechanisms of each method, and they distinguish among four kinds of results produced by various dating methods. Hole and Heizer (1973:247) make a simpler distinction when they identify four bases on which all dating methods stand. This distinction will serve our purposes here. First, recalling time's cycle, time can be measured on the basis of cyclical events, such as the movements of the sun and moon; such events form the basis for modern as well as ancient calendars. Second, time can be based on certain constants such as the speed of light (light years) or the vibration rate of quartz crystals. Third, recalling time's arrow, time can be based on successive and/or cumulative changes in something, for example, in the decay of radioactive substances such as ^{14}C or in changes in artifact form. Fourth, time can be based on the stratigraphic observation of superposed phenomena.

These are very different bases for keeping track of time, and each measures time differently. Calendars and some clocks (sundials, for example) are based solely on observed periodicity in natural events. Calendars present time as elapsed time that is measured relative to the movement of celestial bodies. The modern world divides calendars into various units—years, months, weeks, and days—and uses clocks and watches, the mechanical components of which have been calibrated against standards, to further subdivide days into hours, minutes, and seconds. Some modern digital watches even present time in terms of all seven divisions. The important point is that throughout history, the measurement of time has been based on the repetition of events at uniform intervals. This is not to say that all cultures conceive of time in the same way or even that all cultures have conceived of time itself [see Bailey (1983) and Toulmin and Goodfield (1965) for an anthropological and a historical perspective, respectively]. All it means is that those who have attempted to measure it have sought out regularities, whether it be the revolution of the earth around the sun, the revolution of the moon around the earth, the rotation of the earth, or the vibration rate of quartz crystals.

Humans are not the only organisms that rely on natural phenomena to mark time. Following winter dormancy, trees growing in temperate areas form new growth cells, which show up as incremental rings, the end result being an expansion of the tree's diameter. The process is repeated year after year until the tree dies. Thus a cross-section through a tree can be read in calendarlike fashion to gauge its age (Fig. 1.2). The cross-section is a calendar in the same sense as the one made of paper that hangs on the wall at home, since they are based on exactly the same principle: natural events. The only difference is that the wall calendar is tied to a commonly accepted (historical, see below) point of origin, whereas the tree ring calendar is not.

We commonly think of dendrochronology and radiocarbon dating in the same breath, placing both under the category of absolute dating methods, but each measures a different kind of time because each measures time differently. As opposed to dendrochronology, radiometric dating is based not on the recurrence of natural phenomena but rather on irreversible changes in natural phenomena, changes that can be measured and then converted by way of an algorithm into a date. The last part of the preceding sentence is the most important one: Measured change, derived from the difference between a constant modern ratio of $^{14}C:^{12}C$ and the observed ratio in the sample, is converted into a date. Radiocarbon dating is based on the fact that the unstable isotope ^{14}C decays at a known and more or less constant rate, and thus the $^{14}C:^{12}C$ ratio will decrease over time; this comprises the algorithm. Knowing the rate and the amount of the isotope present in a sample (based on how many beta particles are emitted per unit of time) allows one to calculate, within measurement error, how many radiocarbon years have elapsed since the organism yielding the sample died.

Radiocarbon dates do not match calendrical dates, a phenomenon recognized when wood taken from dendrochronologically dated trees was subjected to radiocarbon dating. It now is known that the atmospheric ratio between the carbon isotopes ^{12}C and ^{14}C has not remained constant over time. Once that was understood, conversion tables were constructed so that one could move back and forth between radiocarbon dates and dendrochronological dates. This sounds simple enough until one realizes that only a few years ago, more than a dozen such "dendrocorrection" tables existed, each differing in significant ways from the others. The Twelfth International Radiocarbon Conference, held in 1985, helped to alleviate some of the confusion by providing a forum in which many of the previous differences were settled (e.g., Stuiver and Becker, 1986). Out of this meeting appeared a special issue of the journal *Radiocarbon* (Stuiver and Kra,

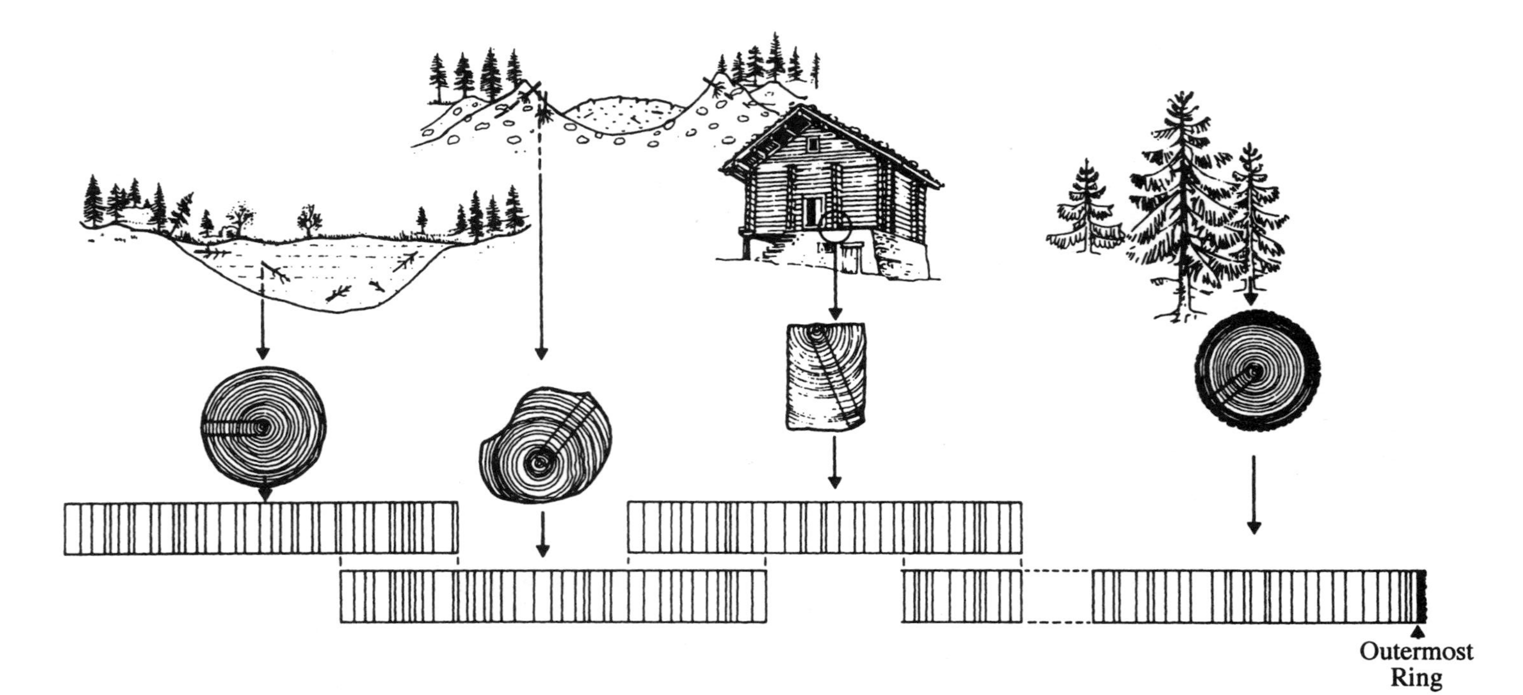

Figure 1.2. A schematic representation of the dendrochronological method. The irregular occurrence of wide and narrow tree rings enables archaeological samples to be dated. Matching the inner layers of living trees with the outer layers of beams in a building allows samples of known and unknown ages to be arranged in chronological order. The addition of more samples of unknown age then can be added to the sequence, provided the samples overlap with the master sequence (after Schweingruber, 1989).

1986) that presented in detail the most recent dendrochronologically corrected age calibrations for radiocarbon dates. A computer program, which originally appeared in the same issue of *Radiocarbon* (Stuiver and Reimer, 1986) is now in an even later release (Stuiver and Reimer, 1993) and is available from the Quaternary Isotope Laboratory at the University of Washington. It has become the standard for archaeology.

Let us ignore the technicalities of correctly reading tree rings or of counting precisely the number of beta emissions from a radiocarbon sample. The more important issue is that there are different ways of measuring time. Earlier we briefly introduced three scales of measurement. Following common use in statistics (Stevens, 1946) there actually are four "scales" of measurement. The term "scales of measurement" is meant to show that there is a rank order in terms of how information-laden each kind of measurement is relative to the others. Each scale is quite appropriate for certain kinds of analysis, and our order of presentation should not be read as meaning that one kind is always superior to the others. Rather, the order we follow is the traditional one of beginning with the scale that contains the least information and progressing to the next scale, which contains more information.

Nominal Scale Measurement

A measurement that plays an important role in everyday life is *nominal* measurement, or literally the naming of categories. Blue marbles are placed in one category and red marbles in another, categories that are equal in rank. The same is true with related things such as blood types and completely unrelated things such as chairs and apples. These are all nominal categories. We even have nominal scale temporal units that we might refer to as time periods, such as "Christmastime" and "the Thanksgiving holidays." These are nominal categories, and the problem is that nominal scale categories are completely unranked. But the Thanksgiving holidays and Christmastime can be ranked in terms of chronological order if they are related to when a calendar year begins. Only by noting such a relation can the Thanksgiving holidays always be chronologically earlier than Christmastime within our calendric year. The folk categories "good times" and "bad times" also can be rank ordered, but in this case rank is based on some scale reflecting the quality of life, not time. Our point is that nominal scale measurements play little or no role in measuring time. To have such a role requires units such as Christmastime and bad times to be ranked relative to some scale, which makes the units not of nominal scale but minimally of ordinal scale.

Ordinal Scale Measurement

There are three kinds of units that measure time in ordered fashion, the first of which are *ordinal* units. Such units are rank ordered in terms of their suspected chronological relation to each other. In some cases we know that one unit comes before or after another, but we do not know how much earlier or later one unit is than another. We might say that in a stratified archaeological site, one where individual layers, or strata, are visible, stratum A was deposited prior to stratum B, but we cannot be more specific in terms of how much time separates the two depositional events. One hundred years might have elapsed between the deposition of stratum A and stratum B, or 500 or even 1000 years might have elapsed. Note as well that we cannot specify how long it took for stratum A to be deposited, nor can we specify the duration of time over which stratum B was deposited. We simply do not know what the rate of deposition was, nor whether that rate was constant. In other cases we may be able to order phenomena into what we think is a chronological sequence, but we cannot determine the direction of the flight of time's arrow based on the criteria used to create the ordering. Such cases occur when seriation is the method of measurement.

Interval Scale Measurement

A second way to measure time in ordered fashion is through the use of *interval* scale units, which are ordinal scale categories that have one additional property: Either there is an equal length of time between events or the categories themselves are of equal length. An example of the first is our concept of midnight, which by definition occurs exactly every 24 hours as measured from the previous midnight. An example of the second is a day, which for most practical purposes always contains 24 hours. What about months? Given that they are of unequal length, do they qualify as interval scale categories? Under the definition we presented above, they do not. Months, however, are not important today except that, based as they once were on a lunar cycle, they are used to determine the date of religious events such as Easter. Otherwise, they serve as a convenience. For example, April 15 is easier to recognize than the 105th day of a nonleap year. As opposed to months, hours and days are consistent in terms of the amount of time each encompasses; they are interval scale units.

Interval scale time is an interesting concept in that it has no true zero point. All civilizations—Mayan, Chinese, Egyptian, Sumerian, for example—had calendars based on the movement of celestial bodies, so their days and years were more or less of the same length. However, they started their calendars at different points, usually based on a religious or mythical event. The Maya calculated their dates as so many days since the end of the last Great Cycle, just as our Gregorian calendar calculates dates as so many years before or after the birth of Christ. These calendars affix an arbitrary zero point to time measured in such a manner; you occasionally will see the term "0 BC" on chronological charts, but such usage is nonsensical because it does not exist on the Gregorian calendar, which comprises the system we use to chart the passage of time. The problem stems from the work of the sixth-century monk Dionysius Exiguus, who was asked by Pope St. John I to compute the date of the next Easter. Dionysius made the calculation, but he did several other things as well, one of which was to compute the date of the birth of Christ. He set the day as December 25 and the year as the 753rd year after the founding of Rome. He then took the Julian calendar, which Julius Caesar had established in 45 BC, and reset it not to Christ's birth but to the day of his circumcision, which traditionally occurred on the eighth day of a Jewish male child's life. That particular day for Christ was January 1, and he called the year in which it occurred AD (Anno Domini) 1; literally, the first year *after* the birth of Christ. The year of his birth was 1 BC, as he was circumcised on the first day of the following year; if the BC–AD clock is reset with his circumcision, then the year before that event must be 1 BC. Thus there was no zero year; hence there can be no such thing as 0 BC or AD 0, unless by those terms we are speaking strictly of a point in time and not a yearlong span.

Regardless, using the instant of Christ's circumcision as a zero point implies nothing about the reality of that zero. In fact, it is not real; time did not start with his circumcision, and this has a major ramification for chronology. Since the zero point is placed arbitrarily along time's continuum, any ratios that we compute are meaningless. Consider the ratio that exists between the number 1000 and the number 500, which is 2:1. Likewise, the ratio that exists between the number 2000 and the number 1000 is 2:1. The fact that the whole numbers are relative to (a natural) zero point allows us to carry out the arithmetic operands of multiplication and division, and thus to compute meaningful ratios. For example, the number 1000 is twice the number 500. Now, consider what happens when we try to compute a ratio using the years AD 1000 and AD 500. One might think that the ratio of those two dates tells us that, relative to the zero point, AD 1000 is twice as old as AD 500, but such a ratio is a logical absurdity. The appearance of a ratio comes from using Christ's circumcision as an arbitrary zero point.

Ratio Scale Measurement

A conceptual zero point and thoughtful choice of a ratio for calculation can avoid such problems. Consider, as Lyman and Fox (1989:312) do in their treatment of bone weathering, taphonomic time, calculated as years since an animal died. Thought of in this way, time becomes a ratio scale measurement because the zero point is fixed at the date of the animal's death. The zero point varies with each individual animal; an animal that died 10 years ago has been dead exactly twice as long as one that died 5 years ago. The ratio of 2:1 years here makes sense given the conceptual zero point of death and its movement across the temporal continuum. Similarly, for the purpose of calculating age, we set the zero point of a person's life at the second he or she is born. Birthdays commemorate this zero point. A baby on its second birthday, literally, the second anniversary commemorating his birth, is twice as old as he was a year ago. This only works because there is a true zero point set at the start of the phenomenon that interests us: the death of an animal or the birth of a baby. Remember that time is a dimension. When we specify the extent of a spatial dimension, we always specify, explicitly or implicitly, the limits of the measurement. Explicitly, it is 50 miles from Mudville to Pottsville; we specify the limits. Implicitly, a race track is 440 yards long; we imply the limits with the word "long," from one end to the other. We must do the same with time.

Does time itself have a true zero point, or has time always existed? The answer to this question is found in the fact that time is a dimension. This was first enunciated mathematically by Albert Einstein with his publication in 1905 of the special theory and in 1915 of the general theory of relativity. Implicit in this work was the consequence that space and time were mathematically the same. Time was a dimension along with the three spatial dimensions. Einstein, however, had inserted an error in his theory before publication. In developing his mathematics, he found that they required a universe that was either expanding or contracting, an impossibility to anyone who believed strongly, as he and most of his contemporaries did, in a Newtonian universe of unchanging and infinite stability. To "correct" this abnormality he inserted a factor, his cosmological constant, that canceled the unwanted result. Local variations were allowed, but the overall conditions of the universe remained constant.

The theories were confirmed by astronomical observations and were rapidly adopted without, it appears, any major questions or reservations. However, in 1927, astronomer Georges Lemaitre, doubting the validity of the cosmological constant, proposed that the universe was expanding. He reasoned that if this expansion were tracked back through time, one would arrive at a point far in the past when the universe was at its smallest possible size. Einstein reacted vigorously, questioning Lemaitre's abilities and denying that his proposal was possible. Quite simply, the universe was infinite and unchanging. Several years later, confronted with Edwin Hubble's astronomical evidence that the universe was expanding, Einstein reportedly admitted that the addition of the cosmological constant was the biggest blunder of his life.

The universe did start small, about 15 billion years ago, with what popularly is termed the "big bang." Opinions vary on the details, but there is general agreement that a small second after the start, perhaps in as little as 10^{-43} second, time separated from other dimensions and became a dimension in its own right. This was the start of time, the zero time from which all future time is measured. Before the big bang, there was no time. This concept of no time is probably impossible to fully comprehend, but an example of a more common measurement might bring the point closer to our understanding. Temperature as a measure of heat usually is expressed as degrees Fahrenheit, a scale where zero is 32°F below the freezing point of water, or as degrees Centigrade, where zero is at the freezing point. Both zero points are arbitrary; the true zero point is 0° Kelvin (approximately −459.67°F and $-273.15°C$), the temperature at which all matter possesses minimal energy. No such thing as temperature exists below 0° Kelvin.

BEYOND MEASUREMENT SCALE: IDEATIONAL AND EMPIRICAL UNITS

No one, archaeologist or other, can help but view time as a continuum. We do not have to be physicists in order to understand that time has existed over a long, continuous span, nor is it difficult to understand that although time is a continuum, it can be broken down into segments—years, days, hours, or whatever—to allow us to perform analytical work. We learned in an elementary geometry class that lines are continuous (within the limits of space time) and that they can be partitioned into segments that have specified beginning and ending points. The same is true of time. As we noted earlier, time exhibits cycles, those repetitions that keep time orderly and allow us both to predict the future behavior of phenomena and retrodict past behaviors based on observations made in the here and now.

But is time real? Yes, it is. We cannot feel or see it, but not all real things have qualities that allow us to directly sense them. However, simply because time is real by no means implies that the *units* we use to categorize time are real. In fact those units are not and never can be real. Rather, they simply are ideational units that we have created to carve up a continuum in ways that make sense to us analytically. The various "ways that make sense" are conditioned by the theory under which one operates. Thus time can be continuous, cyclical, or even discontinuous, depending on the units used to measure it. Regardless of what our individual scientific interests might be, theory specifies the kinds of units to use. We might decide that the color of a stone tool probably is not related to how that tool functioned in the past, whereas the angle of the working edge probably is. Thus if we are interested in functional variation in stone tools, we choose attributes such as edge angles, traces of use wear, and other attributes that theoretically are causally related to the property of analytical interest.

Because we decide on the attributes to record, they are ideational categories, meaning that they are not real in the sense that they can be seen or picked up and held. An edge angle, itself an ideational unit with different empirical manifestations, such as a particular 45° edge, is measured in ideational units known as degrees. Ideational units can be viewed as tools used to measure or characterize real objects. An inch and a centimeter are used to measure length, grams and ounces are used to measure weight. Inches, centimeters, grams, and ounces do not exist empirically; they are units used as analytical tools to measure properties of empirical units, defined as phenomenological things that have an objective existence. A pencil is an empirical unit just as time is. Both are real (empirical) entities that can be measured. But the units we use to subdivide time into segments—centuries, years, hours, nanoseconds, and the like—are ideational units. Keeping one's ideational units separate from empirical units, which the former are meant to measure, is critical to any study of change. Unfortunately, this has not always been made clear in Americanist archaeology. As simple as the distinction might seem, the discipline is a graveyard of arguments about the reality of units used to examine time, an entity that in archaeology usually is equated with culture change.

Science is one sense-making system that requires precision; we cannot leave terms undefined, nor can we depend on inner feelings to get the job done. Constructing ideational units to measure the passage of time archaeologically is much more difficult than simply using ideational units such as days and years to measure it. This is because the units we create and use are inferential, meaning that in and of themselves they do not measure time. Rather, we infer that they are measuring time. Thus we had better be sure that we are using units of the appropriate kind. We examine this important topic in the following chapter.

2 The Creation of Archaeological Types

Archaeological types are, as David Hurst Thomas (1998:235) put it, the discipline's "basic unit[s] of classification.... [They] are idealized categories artificially created by the archaeologist to make sense of past material culture." Archaeologists create types because in dealing with large sets of materials, it is impractical to describe each specimen in excruciating detail. Nor is there necessarily any reason to do so if we can somehow construct groups of things and describe each group in a way that does justice to the things in the groups. By that we mean the description of the group becomes a legitimate proxy for individual specimens in the group. Types indeed are idealized categories, often formed around central tendencies evident in a set of objects. When we state that a particular object is of type A, what we really are saying is that there is a category known as type A into which are placed certain objects that share a suite of particular characteristics. Further, we are implying that specimens placed in that type are more like each other than are specimens not placed in that type.

Rarely will the specimens in type A be clones, meaning that they will differ in certain characteristics. This is not too important if those characteristics are not the distinguishing characteristics, or attributes, that together act as the basis for the type. Oftentimes we will read a type description that says something like "objects placed in type A exhibit characteristics 1, 2, and 3. Further, length usually varies from *w* to *x* centimeters and width from *a* to *b* centimeters." In other words, attributes 1–3 should be present for an object to be placed in type A, but length and width are not too important in making an assignment. This does not mean that variations in length and width are unimportant; all it means is that they are unimportant when deciding whether or not an object falls in a particular type.

No one system of dividing a pile of archaeological specimens into types is necessarily superior to another, because the system and the units contained in it are (or should be) tied to the purpose for which the pile was subdivided in the first place. As Thomas (1998:235) put it, "Artifact types come in all shapes and varieties, and the naked word *type* must never be applied without an appropriate modifier. One must always describe precisely which type of type one is discussing." For example, if our intended purpose is to examine artifact function, we require a different set of characters or attributes than if we are interested in how people decorate objects. Knowing that a pot was used for cooking does not tell us anything about whether it was decorated, just as knowing that it carries incised chevrons on the rim tells us nothing about its function. To answer questions of function, we create *functional* types; to answer questions about decoration we use *decorative* types. We can also create types based on differences in shape-related characteristics and call them *morphological* types. When types are mixtures of decorative and morphological characteristics, we refer to them as *descriptive* types.

There is yet another kind of type—the *chronological* type—and it is the one of greatest concern here. Chronological types, also referred to as historical or temporal types, are, as the names imply, types that are useful for measuring time. As Julian Steward (1954:54) noted, what he termed an "historical-index" type "has chronological, not cultural, significance. It is a time-marker. [It is] used to distinguish chronological ... differences." We indicated in Chapter 1 that time takes on the appearance of a continuum, a cycle, or a discontinuum, depending on the units used to measure it. Beginning in the second decade of the twentieth century, Americanist archaeologists used variation in artifact form as the measuring device; by form we mean any formal property such as shape (the commonsense meaning of form), color, texture, presence–absence of decoration, kind of decoration, and all other properties intrinsic to the artifact. The formal variation of artifacts was and still is recorded as artifact types. Chronological types form the heart of the relative dating methods that are the subject of this book, and thus it is critical to review how such types are constructed. If types are constructed in particular ways, they successfully measure time as a continuum and can be considered to comprise chronological types. But if they are constructed in other ways, they may measure time cyclically, discontinuously, or not at all.

INITIAL CONSIDERATIONS

In Chapter 1 we stated that archaeological dating involves determination of the age or date of "events." What exactly are those events? Albert C. Spaulding (1960:448) indicated that the "prehistoric events usually thought of as archaeologically significant—the targets of chronological inference—are the manufacture and primary deposition of the artifact." He went on to state, "In most instances no serious question is asked about the time interval separating these two events [manufacture and primary deposition]; the general uncertainties of chronological scaling are such that the interval can be treated as negligible without serious difficulty" (Spaulding, 1960:448). Others might disagree with the last statement, noting for example that the duration of time between the manufacture or production of an artifact and its final deposition may vary from instance to instance within a category or type of artifact (e.g., David, 1972; Kristiansen, 1985). Variation in the duration of the "circulation phase" (Kristiansen, 1985:254), or "life span" (David,

1972), of particular artifacts may be an important bit of information for answering some sorts of questions, but such information tells us little about when an artifact was made or when it was deposited. Knowing exactly when something was made (say, to the nearest day) and precisely when it was deposited will provide duration of the use life; one merely subtracts the former from the latter date. In many cases, as Spaulding notes, such is not possible, even today, given the chronological sensitivity of the techniques and units we use to measure time.

If time is measured as a "sensed sequence of events" (Spaulding, 1960:447), what are the events that we might sense or perceive? Irving Rouse (1967:165) provides a long list of such events, but we need not repeat them all here. Rather, given our focus on relative dating and that ordinal scale time is often measured with units generally termed "historical types" of artifacts, we begin with a brief consideration of the sorts of events represented by artifacts before turning to a consideration of what is meant by the term "historical type." We might wish to know when the "fabrication" (Rouse, 1967:165), or "production" (Kristiansen, 1985:254), of an artifact took place. In this case, we want to know when along the temporal continuum the set of attributes comprising a specimen came together, whether through the reduction of a piece of stone to produce a projectile point or the shaping and firing of clay to produce a serviceable vessel. Each individual attribute may be found over a different portion of the temporal continuum; their co-occurrence on a discrete object termed an artifact represents that part of the continuum during which all the attributes considered were combined.

We might want to know when a set of different kinds of artifacts came together to form what are variously termed tool kits, assemblages, or components. Such aggregates of discrete objects generally comprise some sort of depositional event that resulted in the spatial association or combination of the particular artifacts. Because they represent multiple acts of fabrication, aggregates of artifacts tend to span longer durations than a single artifact that represents a single fabrication event. Whether or not we can in fact detect such differences in duration depends on a host of variables that influence the sensitivity and resolution power of the dating technique employed. What is most important to realize is that whether we are dating the fabrication of an artifact or the creation of an aggregate of artifacts, we are dealing with events that occurred over some span of time of greater or lesser magnitude. What we end up with, then, is what James A. Ford (1949:51) termed a "mean" age for that event (by "age" we mean some span of time, not a "date" precise to the nearest calendar year; see Chapter 1). Do not take the term "mean" literally; what is meant is perhaps better expressed as a "modal" age, given how we have considered the formation of the material being dated.

If the target "event" is the manufacture of a projectile point, then we must realize that that event took place over some span of time, even if only an hour or two, and the attributes of the point were created in some sort of sequence across that span. Although we might suspect what the precise sequence was, it is doubtful we will know it in detail for the point in hand. Given that the point comprises attributes, if attribute 1 was created during time span A–D, attribute 2 during time span A–E, and attribute 3 during time span C–G, then the modal age would be time span C–D. Similarly, if we take as our target event the deposition of an aggregate of artifacts, we first consider that each artifact within the aggregate was made during a particular span of time (assuming they are all more or less historical types). If one was made during time span A–C, another during time span B–D, and a third during time span B–E, then the modal span is B–C, and that comprises the modal age of the assemblage. The same applies if the particular time spans concern when the artifacts were deposited rather than when they were made.

CONSTRUCTING CHRONOLOGICAL TYPES

How do we construct chronological types? The answer is fairly straightforward: Determine the characteristics or attributes of objects that change over time. These could be attributes of shape, such as the location of notching on projectile points, or they could be attributes of decoration, such as the colors applied to the surfaces of ceramic vessels or the designs displayed by the colors. Once such characters have been identified, use them as the basis for defining chronological types. In other words, chronological types are constructed by selecting attributes that either themselves change through time or the combinations of which change through time. How do we know which attributes or combinations thereof are useful? To be useful, chronological types must have what Alex Krieger (1944:272) called "demonstrable historical meaning." An example from the early culture historical period will illustrate what Krieger meant.

Nels Nelson was a Danish immigrant who eventually found his way to the University of California and became a student of A. L. Kroeber, who in turn had been a student of Franz Boas at Columbia University. Nelson subsequently was hired by Clark Wissler of the American Museum of Natural History to begin a field project in the Galisteo Basin of New Mexico, just south of Santa Fe. He spent 1911–1912 in New Mexico, returning to the region in 1914–1915. Figure 2.1 illustrates four types of pottery Nelson identified based on his stratigraphic excavations at Pueblo San Cristobal and other sites in the Tano ruins district in and around Santa Fe in 1912 and 1914 (Nelson, 1916). Nelson segregated his pottery into seven types: the four shown in Fig. 2.1 plus three others. He relied primarily on surface treatment as the basis for creating the types and referred to them in terms of those treatments. One he termed "biscuit ware," calling it a "peculiar kind of pottery, which can be detected even by the touch," and another "corrugated or coiled ware," which "is almost invariably covered with soot and was evidently made exclusively for cooking purposes" (Nelson, 1916:168). Type I was referred to as "two- and three-color painted ware," type II as "two-color glazed ware," type III as "three-color glazed ware," type IV as "historic two-color glazed ware," and type V as "modern painted pottery."

Figure 2.1. Four types of pottery created by Nels C. Nelson in 1916 based on his excavations in the Galisteo Basin of New Mexico (from Nelson, 1916) (photograph courtesy of the American Museum of Natural History).

Notice that in his categorization Nelson made use of several kinds, or what we refer to as *dimensions*, of variation. One had to do with what we refer to as decoration: the number of colors painted on a vessel; another had to do with the initial stage of vessel manufacturing: biscuit ware and corrugated ware (we doubt corrugation of a vessel's surface was decorative); and a third could be decorative or not: the presence of glazing. At first glance, we might propose that some of the characteristics Nelson used, such as corrugation, are related to how a vessel functions. Other characteristics, such as how many colors a vessel exhibits, probably have no effect on function. Glaze might affect the function of a vessel, but it certainly keeps paint from peeling and thus ruining the design. The bottom line is that although Nelson mixed and matched decorative and what we might term "technofunctional" vessel characteristics, the resulting types formed homogeneous groups, meaning that vessels and sherds within any one type bore more resemblance to each other than they did to specimens in other types. Could he have constructed different types using the same pottery assemblages? Yes, and in fact Nelson freely admitted that his classification was "no doubt arbitrary, but it will serve present purposes" (Nelson, 1916:168). The classification did indeed serve "present purposes," which was to establish a way to measure time and thereby produce a chronological ordering of archaeological materials in the north-central Rio Grande region of New Mexico, and it did it remarkably well.

Did Nelson simply get lucky in that his types measured time? If he had created other types instead of the ones he did, would they also have been good chronological types? Maybe, but the point is that the ones he did create were not simply the products of luck. They were based on observations, both stratigraphic and otherwise, that Nelson and others made in the Pueblo region of New Mexico. Might someone argue that it is tautological to use stratigraphic observations to construct types that then will be used to measure the passage of time? If Nelson or someone else had never tested his types to determine their chronological reliability outside the site where the stratigraphic observations were made, that would be not only tautological but also bad science. But Nelson did check the sequence of types. It is clear in his excavation report (Nelson, 1916) that before he began excavating at San Cristobal he suspected, based on stratigraphic evidence from other sites, what the chronological order was of four of the types, including corrugated ware and biscuit ware. A fifth pottery type obviously was from the historical period, since it was found consistently with bones of the horse and other historically introduced domestic animals. But Nelson needed more evidence than what suspicion afforded, and thus he tested the suspected sequence of types first at Paaco, a ruin just northeast of Albuquerque, and then at San Cristobal, where he found the most complete sequence.

By the time he finished the excavations at San Cristobal, Nelson had found tangible proof of the chronological position of the types. His pottery types were of the temporal sort, which means they had "demonstrable historical meaning"

(Krieger, 1944:272): They came into being, gained in popularity to a maximum, and then began to fade, finally disappearing. This waxing and waning of pottery types is illustrated in Fig. 2.2. To be of chronological use, types, rendered as particular combinations of attributes, should have continuous distributions in time, and the period of time over which they occur should be fairly short. In other words, each type should have occurred only once, and it should have disappeared after a relatively short life (Fig. 2.2). That is, chronologically useful types cannot "reappear" at a later date. Time's arrow is of interest, not time's cycle in this case, which is not to say the latter will not be of interest given an analytical problem other than marking the passage of time. Further, when we plot the relative frequency of a type through time, we should obtain a reasonably close approximation of a unimodal curve. This kind of temporal distribution, as we will see in Chapter 4, is the principle behind frequency seriation. If we get a discontinuous distribution or a multimodal curve over time, then the type as constructed is useless for chronological purposes.

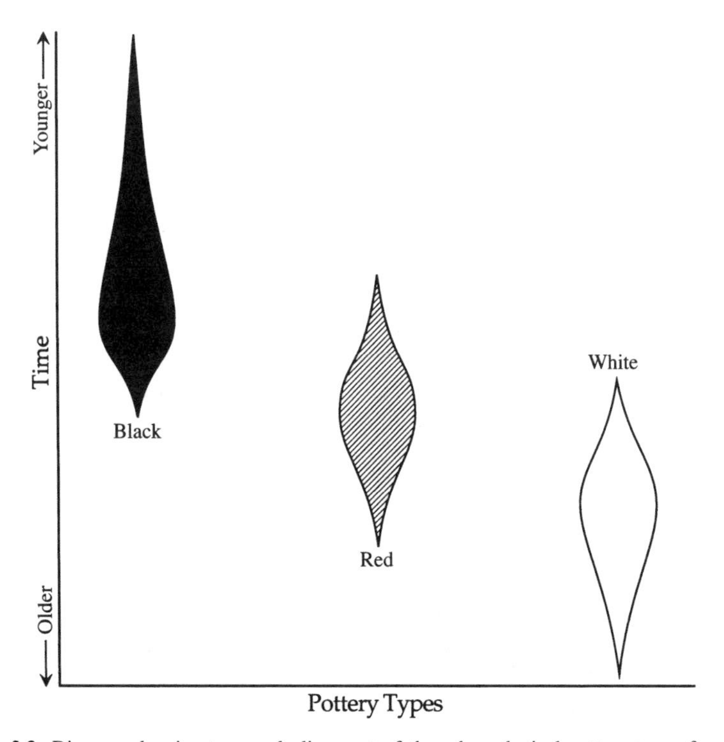

Figure 2.2. Diagram showing temporal alignment of three hypothetical pottery types from latest (black) to earliest (white).

Pottery is not the only kind of artifact that lends itself to the creation of chronological types. If it were, much of the archaeological record of North America would be undatable except through the use of radiometric methods. Pottery production may have begun in the Southeast as early as the third millennium BC (Sassaman, 1993) and in the Midwest as early as 1100 BC (Reid, 1984; Skibo *et al.*, 1989), though it was not until 500–600 BC that it became fairly common. Pottery occurred even later in other parts of North America. Projectile points, however, were manufactured throughout the 11,000-plus years of human tenure in North America.

Projectile points do not normally carry what we would characterize as "decoration," but they can be placed into types based on the presence of certain morphological characteristics. Figure 2.3 illustrates several projectile points, similar specimens of which occur on archaeological sites in the central Mississippi River valley. No one can keep track of tens of thousands of specimens, but a few dozen kinds are not so difficult to remember. Several types have been created to subdivide the tremendous variation in projectile points that were manufactured in the Mississippi Valley between roughly 9250 BC and AD 1700 (all dates are in uncalibrated radiocarbon years). Some types are described so specifically that there is little variability allowed between the specimens placed in them, whereas other types are more loosely constructed and encompass more variation.

The points shown in Fig. 2.3 date between roughly 8000 BC and 7000 BC. We say "roughly" because radiocarbon dates for sites producing these early points are rare. Further, there is considerable evidence (O'Brien and Wood, 1998) that the few dates that we do have are problematic. In brief, there is nothing wrong with the dates themselves but rather with the association between the radiocarbon samples and the projectile points they purportedly date, a typical difficulty of indirect dating methods. At present we cannot say for sure whether points of one type preceded those of another or whether all the points were manufactured at roughly the same time. Notice the similarities among the specimens, despite the fact that they commonly are placed in three separate types. All are rather long and have fairly broad blades, and all have notches. Although it does not show on the figures, the edges of the hafting elements are usually ground in order to dull them so that they do not cut the bindings that hold the points to the shafts or foreshafts. Division of the specimens into types is based primarily on whether the notches occur on the sides, the corners, or the base, though notice on the examples of the Thebes type (Fig. 2.3, top row) that two specimens exhibit corner notches and a third specimen side notches. Some archaeologists would go so far as to subdivide the Thebes type into two types based on the shape of the notches, which can be circular or E-shaped, as on the righthand specimen. In fact, Gregory Perino (1985), from whose projectile point guide we extracted the drawings of the points, creates separate varieties based on notch shape. There is no rule that dictates when it is advisable to create new types. For example, if one is interested in chronological

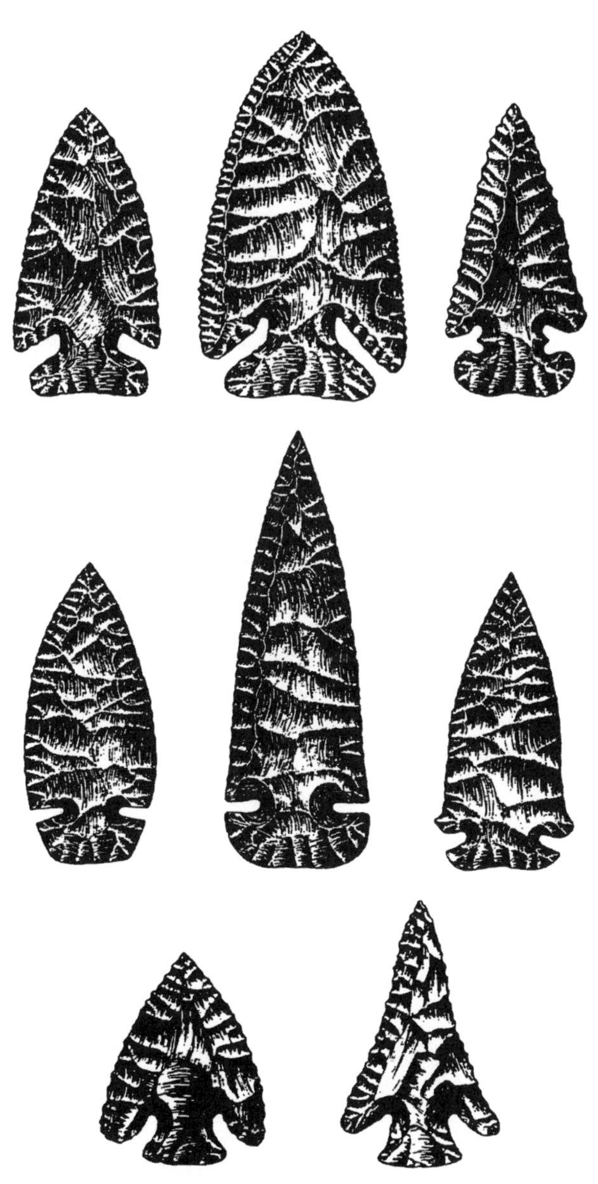

Figure 2.3. Three types of projectile points from the central Mississippi River valley that date ca. 8000–7000 BC: top row, Thebes; middle row, St. Charles; bottom row, Lost Lake. Center specimen in the top row is 11.1 cm long (after Perino, 1985).

types, and it can be demonstrated that notch shape changed over time, then it makes sense to create two types. Again, it all depends on the purpose of the classification and the analytical work to which the types are being put.

CHRONOLOGICAL TYPES IN AMERICANIST ARCHAEOLOGY

Americanist archaeology was not always replete with pottery and projectile point types, and certainly the manner in which types are constructed has changed significantly over the years. In the United States, early classification systems were developed solely as a way to enhance communication between researchers who had multiple specimens they wanted to describe (Dunnell, 1986a). Since the intent of the persons devising the classification schemes was to standardize terminology, most of the systems were based on readily perceived differences and similarities among specimens. This meant that form received the lion's share of attention, though form and function often were conflated, despite the best efforts of the classifiers to keep them separate. As a result, piles of more or less similar-looking specimens created by late nineteenth- and early twentieth-century classifiers lacked any archaeological meaning, and thus were incapable of doing analytical work. Variation in artifact form within each pile, and to some extent between piles, of specimens had no perceived explanatory value and was simply conceived as noise resulting from different levels of skill in manufacturing, from raw material quality, or from individual choice. As Charles Rau (1876:159), who was connected with the US National Museum, observed,

> A classification of the arrowheads with regard to their chronological development is not attempted, and hardly deemed necessary. North American Indians of the same tribe (as, for instance, the Pai-Utes of Southern Utah) arm their arrows with stone points of different forms, the shape of the arrowhead being a matter of individual taste or convenience.

Given such ready-made explanations of the remarkable variation evident in the archaeological record, there was no reason to think that artifacts might be useful for keeping track of time.

Although he did not create or use types per se, William Henry Holmes presaged much of the basis for ceramic typology during his late nineteenth-century analysis of pottery from the central Mississippi River valley. About various technological, stylistic, and functional attributes of pottery manufactured (we now know) during the Mississippian period (post-ca. AD 900) he wrote:

> The material employed was usually a moderately fine-grained clay, tempered, in a great majority of cases, with pulverized shells. The shells used were doubtless obtained from the neighboring rivers. In many of the vessels the particles are large, measuring as much as one-fourth or even one-half of an inch in width, but in the most elegant vases the shell

> has been reduced to a fine powder. Powdered potsherds were also used. The clay was, apparently, often impure or loamy. It was, probably, at times, obtained from recent alluvial deposits of the bayous—the sediment of overflows—as was the potter's clay of the Nile. There is no reason for believing that the finer processes of powdering and levigation were known. A slip or wash of very finely comminuted clay was sometimes applied to the surface of the vessel. The walls of the vessels are often thick and uneven, and are always quite porous, a feature of little or no importance in the storage of drinking-water, but one resulting from accident rather than from design. (Holmes, 1886:372–373)

In that short statement, Holmes noted that (1) some pottery was tempered with finely crushed shell, while other pottery contained larger shell particles (technological and functional attributes); (2) vessel walls often were thick (technological and functional attributes); and (3) vessel surfaces were slipped (technological plus stylistic and/or functional attributes). In other words, his types were composites of different kinds of characteristics in the same way that Nelson's were. Holmes was interested in explaining variation in ceramic vessels, but he was not interested in tracking change over time. Holmes and his colleagues at the Bureau of American Ethnology did not think there had been enough time for significant change to have occurred, and consequently, Holmes "was forced to seek other (e.g., racial, environmental, and diffusionist) explanations for the differences" among the vessels (Meltzer and Dunnell, 1992:xxxvii).

Typological Issues Begin to Take Shape

It would be up to Nelson and his colleagues working in the Southwest, especially A. L. Kroeber and A. V. Kidder, to create pottery types and to demonstrate that some of them measured time. The efforts of Nelson, Kroeber, and Kidder in the teens spawned a cottage industry in terms of Southwestern pottery-type construction that lasted throughout the 1920s and 1930s. We cover several interesting aspects of that industry in Chapter 3, but here our focus is more on the Southeast, specifically the Mississippi River valley, because it was there where many of the conceptual issues surrounding the creation and successful use of chronological types were brought into sharp focus between 1936 and 1951. The two people most responsible for bringing the issues to the forefront were James A. Ford and James B. Griffin. To appreciate the complexities of the issues involved in the archaeological use of types, one can do no better than to read what they had to say on the subject. Their work on typology, work that was methodological as well as practical, directly influenced the creation of archaeological types that are still used to partition the archaeological record of large portions of the central Mississippi Valley.

The initial stages of the typological work were set when Ford undertook a large-scale surface collection of sites in southwestern Mississippi and northeastern Louisiana during the late 1920s and the early 1930s. In the resulting publication,

Ford (1936a:8) noted that, "In the Lower Mississippi Valley, the requirements for a cultural factor suitable for an analysis of time change appear best to be met by the fragments of domestic pottery which have been broken and unintentionally deposited on the sites of the old villages." In other words, broken pottery was plentiful, and it was useful archaeologically because it could be used to perform analytical work, in this case, the measurement of time. Ford used stratigraphic evidence from a site in Louisiana and surface collections of sherds from 103 sites in Louisiana and Mississippi to define seven ceramic complexes, four identified with historical period tribes that occupied the lower Mississippi River valley in the sixteenth and seventeenth centuries—the Tunica, the Caddo, the Choctaw, and the Natchez—and three that were prehistoric—Coles Creek, Deasonville, and Marksville. Ford's use of the term *complex* for the units implied that each unit comprised specimens of a number of ceramic types and other items (O'Brien and Lyman, 1998). Some of Ford's types, which were constructed around decorative characteristics, are shown in Fig. 2.4.

Although Ford made clear his ideas on classification and nomenclature in his unpublished master's thesis (Ford, 1938b), those ideas are better known because of a conference report on Southeastern pottery that Ford coauthored with Griffin (Ford and Griffin, 1938). In both papers, Ford was unequivocal over both the way types had been used by archaeologists up to that point and how types should be used:

> The inadequacy of the procedure of dividing pottery into "types" merely for purposes of describing the material is recognized. This is merely a means of presenting raw data. Types should be classes of material which promise to be useful as tools in interpreting culture history. (Ford and Griffin, 1938:3)

Thus in 1938 Ford established his position, and it would remain essentially unchanged throughout his career: Artifact types are nothing more than tools created and used to order archaeological materials chronologically. His approach to creating types was the same as Nelson and others had used in the Southwest: Use whatever works for measuring the passage of time. This simple guide would come under tremendous attack a decade and a half later.

The work of Ford and Griffin reached a zenith during a large-scale survey of the Mississippi Valley they conducted in collaboration with Philip Phillips (Phillips *et al.*, 1951), the purpose of which was to examine the northern two thirds of what properly is called the Mississippi Alluvial Valley, an area that extends from roughly the mouth of the Ohio River at Cairo, Illinois, to Vicksburg, Mississippi. It is difficult to overestimate the impact that the final volume produced by Phillips, Ford, and Griffin had, not only on the archaeology of the Mississippi River valley but on Americanist archaeology generally (Dunnell, 1985; O'Brien and Dunnell, 1998; O'Brien and Lyman, 1998). In it we find some of the clearest statements ever set forth not only on how archaeological types are created and used but also on the conceptual problems in constructing chronological types.

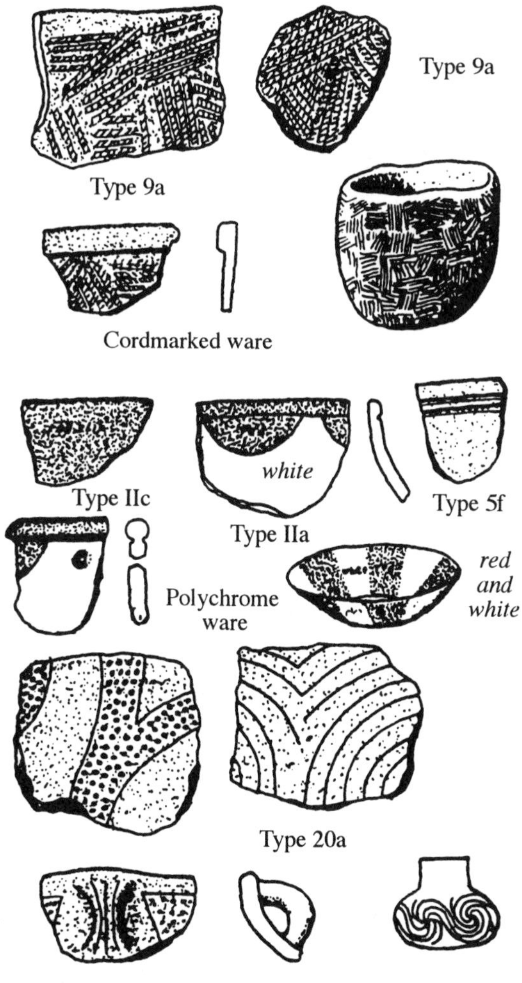

Figure 2.4. James A. Ford's illustration of pottery designs represented in the Deasonville complex of western Mississippi. Some of the designs formed the basis of Ford's marker types (e.g., type 9a) (from Ford, 1935b).

The method adopted by Phillips, Ford, and Griffin for constructing types stemmed directly from Ford's earlier work, as did the rationale behind constructing types: Types were considered to be useful for assigning temporal order to archaeological materials. If they also served another purpose, for example, if the traits used to sort the pottery "correspond to characters that might have served to distinguish one sort of pottery from another in the minds of the people who made and used it" (Phillips *et al.*, 1951:63), so much the better (at least as far as the authors were concerned), but chronological ordering was of primary importance.

Selecting Characters for Building Types

Types were recognized as composites of various characters—paste, vessel form, surface treatment, and decoration (the same dimensions discussed earlier by Holmes)—and Phillips, Ford, and Griffin reasoned that each character had its own history. They viewed it as "unreasonable" to assume that one character was always more useful than another for ordering archaeological phenomena chronologically or to insist that all pottery should be sorted consistently on the basis of the same characters. However, they also recognized that to avoid confusion and miscommunication one should

> select the most sensitive—and at the same time most recognizable—characters as guides or "constants" in the process of classification. In Southeastern pottery generally, these are features of surface treatment and decoration, and thus it has come about that what may for convenience be called the Southeastern classification employs a binomial system of nomenclature in which the second term or "constant" is descriptive of surface treatment or decoration, as in Mulberry Creek *Cord-marked* or Indian Bay *Stamped*. (Phillips *et al.*, 1951:64–65)

How good were the types that Phillips, Ford, and Griffin created? From the standpoint of time, did they conform to what Kidder (1936a:xx) had earlier termed "recognizable nodes of individuality," meaning, did they appear, reach a zenith in terms of frequency (referred to as "popularity" by early culture historians), and then disappear? Nelson's descriptive types did this, which is why they were good chronological types, and so did many of the ceramic types created by Phillips, Ford, and Griffin. For example, they created the type Walls Engraved/Incised (Fig. 2.5) to include pottery from the lower Mississippi Valley that seemed, in the preradiocarbon era, to date late in the archaeological sequence. We now know that Walls Engraved/Incised vessels were indeed late additions to the sequence, first appearing perhaps sometime around AD 1450 and lasting for a century or so. It and numerous other types built around particular kinds of decoration as key identifying characteristics are extremely useful in keeping track of time in the vast archaeological record of the central Mississippi Valley, and the reason they are so useful is that pottery designs, the basis for the types, changed rapidly.

Let us contrast Walls Engraved/Incised with two of the most commonly occurring pottery types on sites in the Mississippi Valley that date after about AD 900: Neeley's Ferry Plain and Bell Plain. The latter are not good chronological markers, unless one's only interest is whether or not an archaeological deposit dates ca. post-AD 900, which was more or less the time when shell-tempered pottery began to be produced widely in the central Mississippi River valley. Both archaeological types were constructed around paste characteristics as opposed to decoration (vessels placed in the two types carry no decoration), and both form the basis for other type descriptions that are based on decoration. By this we mean that in describing some of their decorative types, Phillips, Ford, and Griffin specified

Figure 2.5. Walls Engraved bottle from Pemiscot County, Missouri (13 cm high) (from O'Brien and Fox, 1994a).

that some designs occurred exclusively on vessels of one paste or the other. Thus understanding the difference between what is meant (or implied) by Neeley's Ferry Plain, described as being a coarse paste type, and Bell Plain, described as being a fine paste type, is key to understanding differences among the myriad types and varieties that have been developed to address the variation evident in pottery from the Mississippi Valley. The terms *coarse paste* and *fine paste* are shorthand notations that refer to the degree to which shell employed as temper in vessel paste was crushed prior to its inclusion. Coarse paste refers to a paste that was tempered with larger pieces of shell, and fine paste refers to a paste tempered with shell that was crushed so finely that it often became powder. The difference in pastes can be seen in Fig. 2.6. No shell particles are visible in the top bowl (Bell Plain), but they are clearly visible in the bottom bowl (Neeley's Ferry Plain).

The result of maintaining Neeley's Ferry Plain as an analytical unit is that all coarse-tempered, undecorated body sherds are lumped into a single type. Those sherds could have come from a vessel that, if complete, would have been placed in one of the decoration-based types established for the region, but since the plain sherd was separated from the decorated portion of the vessel through breakage, it is labeled as Neeley's Ferry Plain. Thus that type has become a default category, and as a result corresponding frequencies of the type tend to dominate ceramic assemblages (O'Brien, 1995; O'Brien and Fox, 1994a,b; Phillips, 1970). The same is true for undecorated sherds that come from vessels tempered with finely crushed shell, which are placed in the type Bell Plain. Even worse, Bell Plain, defined as a shell-tempered paste type, has been expanded to include clay-tempered specimens that exhibit the other characteristics of Bell Plain (e.g., smooth interior and exterior finish). Future studies will have to address the problem of mixed pastes and develop a quantitative means of assessing the variation in paste because at present it appears to have significant functional implications; meaning that it had an effect on the usefulness of a vessel and by extension an effect on the lives of the users of the vessel (Dunnell and Feathers, 1991; O'Brien and Holland, 1990, 1992; O'Brien *et al.*, 1994). Note that we say that variation in paste "appears" to have significant functional implications. Simply because a characteristic is related to how a vessel was made or used in no way implies that that characteristic is related to function, by which we mean, in a general sense, the purpose or role of the characteristic in a mechanical or engineering sense. Certain features may be "neutral," meaning that they really do not affect the function of the vessel. This neutrality is what some archaeologists (Dunnell, 1978; O'Brien and Holland, 1990, 1992) refer to as *style*.

It is becoming increasingly apparent that pottery types based on technological and/or functional characters, which the shell tempering is, are apt to be poor chronological markers. This is not to say that all such types are worthless for chronological purposes, because clearly they are not. What we are saying is that types based on stylistic features such as decoration tend to perform their analytical purpose of measuring time better than functional types. The reasons for the superiority of decorative over functional characteristics are complex (see Dunnell, 1978, and O'Brien and Holland, 1990, 1992, for extended discussions), but they translate into the ways that decoration changes over time relative to functional characters. Earlier in this chapter we alluded to the fact that functional types might reappear in a temporal sequence, and thus measure time's cycle rather than time's arrow. Such is certainly important knowledge to have if one is interested in the history of the adaptations evidenced within a cultural lineage—time's cycle—but not if one wishes to determine the ages of the multiple instances of a particular adaptational trait—time's arrow. It is the latter that concerns us here. Similarly, a particular attribute—functional, stylistic, or technological—may reappear more than once

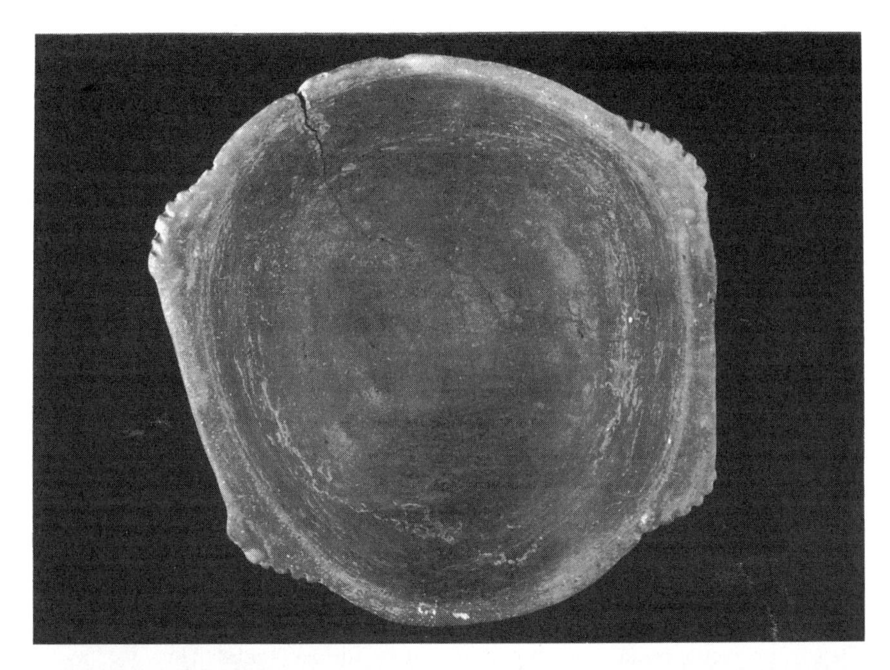

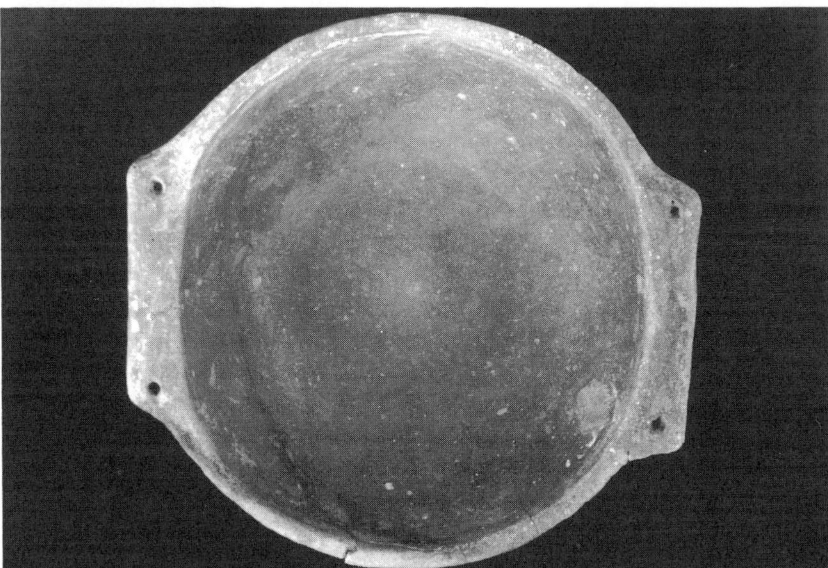

Figure 2.6. Shell-tempered bowls from Pemiscot County, Missouri: top, Bell Plain; bottom, Neeley's Ferry Plain. Bowls are approximately 17 cm wide (from O'Brien and Fox, 1994a).

in a sequence for any of myriad reasons (e.g., Rands, 1961), but again this is a manifestation of time's cycle rather than of time's arrow. For purposes of archaeological dating and measuring time's arrow, we must select attributes or combinations thereof (types) that meet the historical significance criterion. Generally, decorative attributes work better than other kinds of attributes.

The ways in which vessels made by groups living in the Mississippi River valley after ca. AD 1200 were decorated changed dramatically over short periods of time. Vessels manufactured around AD 1200 and those manufactured around AD 1400 might have been decorated either by painting or incising the exterior surfaces, but the designs themselves often were vastly different. This is why the decoration-based types devised by Phillips, Ford, and Griffin are such excellent temporal markers and the paste-based types Neeley's Ferry Plain and Bell Plain, the undecorated types, are not good temporal markers. They might tell us that an assemblage postdates ca. AD 900, but that is about all they tell us. In establishing those types, archaeologists switched emphasis from decoration-based characteristics to functional characteristics. In fact, they had to switch emphasis to distinguish among kinds of vessels that carried no decoration, but this does not mean that the resulting types were useful for measuring the passage of time.

Our argument concerning the superiority of decorative types as chronological tools presupposes that the artifacts in question are capable of being decorated. Clearly, projectile points are not good mediums for decoration. Thus, we have to turn to technological aspects (which we assume are related to how projectile points were made) to build chronological types. Obviously, only those technological aspects that changed relatively quickly are useful. By 1950, the development of chronologically sensitive pottery types had far outdistanced the development of historical types of projectile points in North America. Pottery had long figured prominently in Americanist archaeology, reaching a descriptive peak with the work of Holmes and an early analytical peak with chronological work in the Southwest. As a result, later archaeologists such as Ford and Griffin were preconditioned to regard pottery as integral to the chronological ordering of archaeological deposits. In addition, pottery is abundant on sites across the Southwest and the East, and it was natural that it would assume a position of primacy as an analytical tool. Projectile points, on the other hand, are relatively much rarer occurrences on most sites in these areas. And a sherd, no matter how plain and inelegant, can be placed in a type (if it is not decorated, use a default category such as Neeley's Ferry Plain or Bell Plain), but broken projectile points are more difficult to place.

The first effort to standardize projectile point types on a large, regional scale was *An Introductory Handbook of Texas Archeology* (Suhm *et al.*, 1954), which was reissued under the more appropriate title, *Handbook of Texas Archeology: Type Descriptions* (Suhm and Jelks, 1962). The handbook listed not only the type names of pottery and projectile points from Texas and neighboring regions but also the geographic range of the items and, where known or suspected, the date ranges.

The success of the handbook can be linked directly to its appeal to amateurs and collectors as well as to professionals, an appeal that came from the book's simplicity. Anyone who wanted to know what type of projectile point he or she possessed could open the book and find a similar, named specimen among the photographs. The authors had drastically cut down on the confusion in the literature over what a specimen of a particular shape should be called, a confusion created in part by the proliferation of point type names that was occurring in the late 1940s and early 1950s (e.g., Bell and Hall, 1953; Krieger, 1947), as more archaeologists started creating their own types without searching the literature to see if similar points already had been named. By 1960, at least two other guides to projectile point types had been published by the Oklahoma Anthropological Society (OAS) (Bell, 1958, 1960), and they incorporated established types and added new ones. Those volumes also became best-sellers. They were followed by two more OAS volumes (Perino, 1968, 1971) and by Gregory Perino's (1985, 1991) two volumes entitled *Selected Preforms, Points and Knives of the North American Indians* and Noel Justice's (1987) *Stone Age Spear and Arrow Points of the Midcontinental and Eastern United States.*

Despite the success of such books, the same problems that arose in ceramic typology applied to projectile point typology. Tremendous variation was encompassed within some of the types, so much so that any group of five archaeologists looking at a particular point might place it in five different types. We saw this problem earlier with respect to the specimens in Fig. 2.3. And yet some point types, because of the characteristics used to construct them, became useful chronological markers. Anchoring one end of the North American time scale are large dart or spear points placed in types such as Clovis (9250–8950 BC), Folsom (8950–8650 BC), and Dalton (8950–7900 BC) (Figs. 1.1 and 2.7). Anchoring the later end of the scale are arrow points of dozens of different types, a few of which are shown in Fig. 2.8. In between Clovis at one end and arrow point types at the other were hundreds of other types, some of which also are useful chronological markers.

Where Does One Type End and Another One Start?

Just as selecting the attributes or characters around which to construct a typological system was viewed as arbitrary by Phillips, Ford, and Griffin, so too was the decision about where to "stop" one type and to begin another:

> Each community that had reached a certain level of sophistication in pottery-making will be found to have been maintaining side by side several different vessel styles [read *types*].... Between these centers, styles vary and trend toward those of other centers in rough proportion to the distances involved, subject of course to ethnic distributions and geographic factors. Thus we have in mind the concept of a continuously evolving regional pottery tradition, showing a more or less parallel development in and around a number of centers, each of which employs a number of distinct but related styles, each

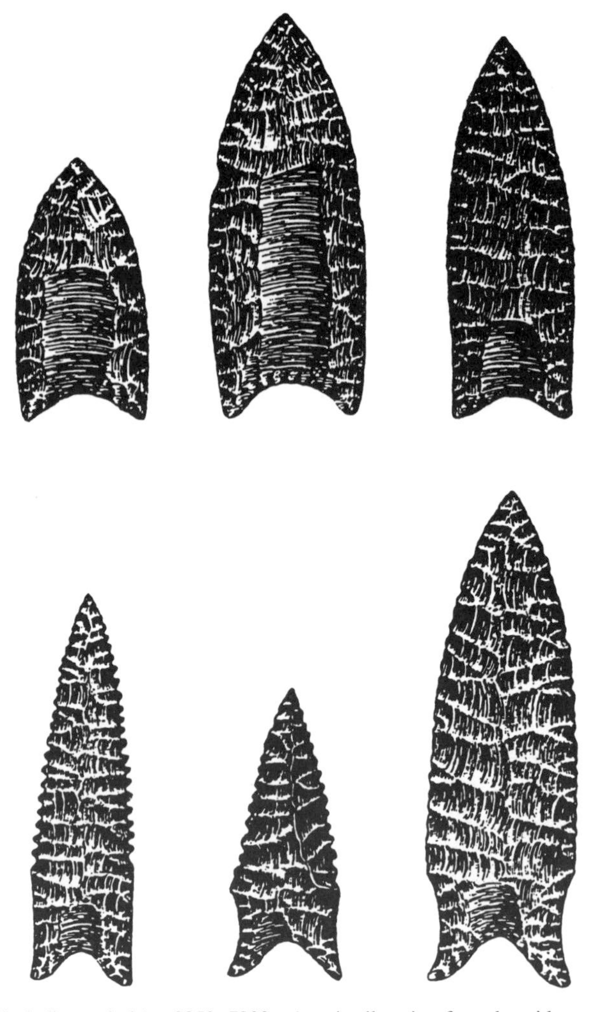

Figure 2.7. Paleoindian period (ca. 9250–7900 BC) projectile points from the midwestern and eastern United States: top row, Clovis; bottom row, Dalton. Center specimen in the top row is 9.1 cm long (after Perino, 1985).

> style in turn being in process of change both areally and temporally. With this remarkably unstable material, we set out to fashion a key to the prehistory of the region. Faced with this three-dimensional flow, which seldom if ever exhibits "natural" segregation, and being obliged to reduce it to some sort of manageable form, we arbitrarily cut it into units. Such *created units of the ceramic continuum* are called *pottery types*. (Phillips *et al.*, 1951:62–63)

Here Phillips, Ford, and Griffin underscored the nature of what we elsewhere (Lyman *et al.*, 1997b) refer to as the materialist perspective; that is, that the form of

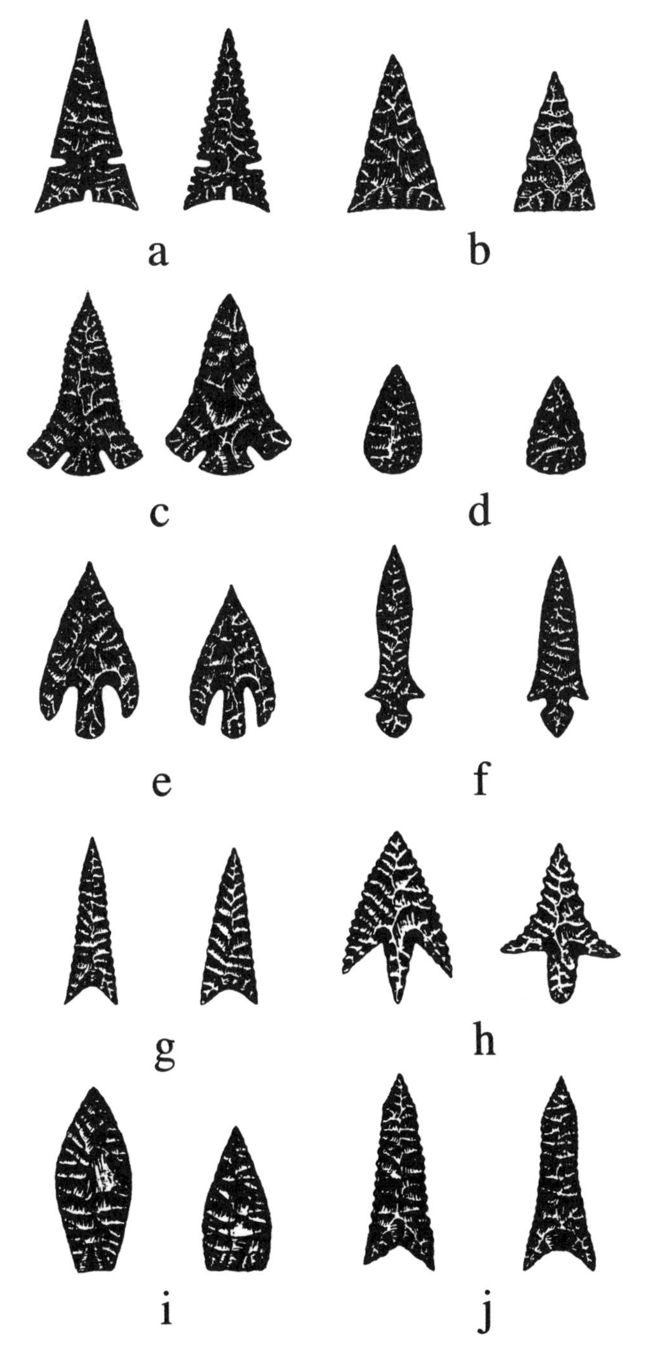

Figure 2.8. Various arrow points from the United States: (a) Cahokia; (b) Caraway; (c) Catahoula; (d) Cottonwood; (e) Deadman's Point; (f) Klickitat; (g) Maud; (h) Perdiz; (i) Shetly; (j) Talco. Upper left specimen is 2.7 cm long (after Perino, 1985).

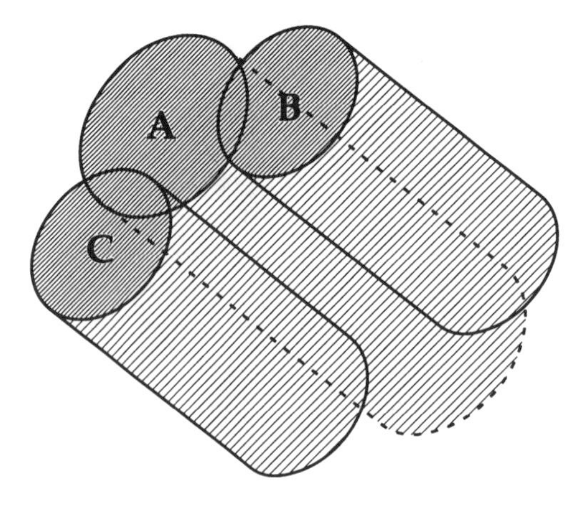

Figure 2.9. Model of the way in which Philip Phillips, James A. Ford, and James B. Griffin conceived of pottery types. Each letter denotes a type, the circle around it denotes the range of formal variation within the type, and the length of the cylinder represents the temporal range of a type (after Phillips *et al.*, 1951).

pottery, like time, is a continuum. Once types were created, they were viewed as having more or less continuous distributions in time and space (Fig. 2.9). In view of what he wrote a decade and a half later, it is ironic that Phillips would buy into the notion that there was such a thing as a ceramic continuum, because he complained bitterly about Ford's (1951) arbitrary method of chopping up that continuum (O'Brien and Lyman, 1998). What is more ironic is that he would lend his name to the statement that pottery types are *created* out of that continuum as opposed to units that can be *discovered* because of natural discontinuities. By 1970, Phillips was totally convinced that artifact types were real, empirical units waiting to be found. As we will see, he was not alone in his convictions.

The Spatial Nature of Artifact Types

Phillips, Ford, and Griffin (1951:426) acknowledged the usefulness of their typological system: "Our classification cannot be too bad or it would not have produced the consistent patterning of types through time," but they also noted dissatisfaction. One of the problems with which they wrestled was what happens to types when space is thrown into the equation. They clearly recognized that

> The groups of ideas to whose products have been tagged such names as Mazique Incised did not spring up simultaneously all over the area. They moved from one part to another, and that took time. For example, the idea of red slipping on clay-tempered vessels (Larto Red Filmed) apparently was moving from south to north through the region, while cord-marking on clay tempered pots (Mulberry Creek Cord-Marked) was moving from northeast to south. Naturally, the former is earlier in the south, and the latter to the north. (Phillips *et al.*, 1951:229)

Figure 2.10 illustrates the spatial distribution of one of their pottery types, Parkin Punctated. Notice that sherds of this type occur most frequently on sites

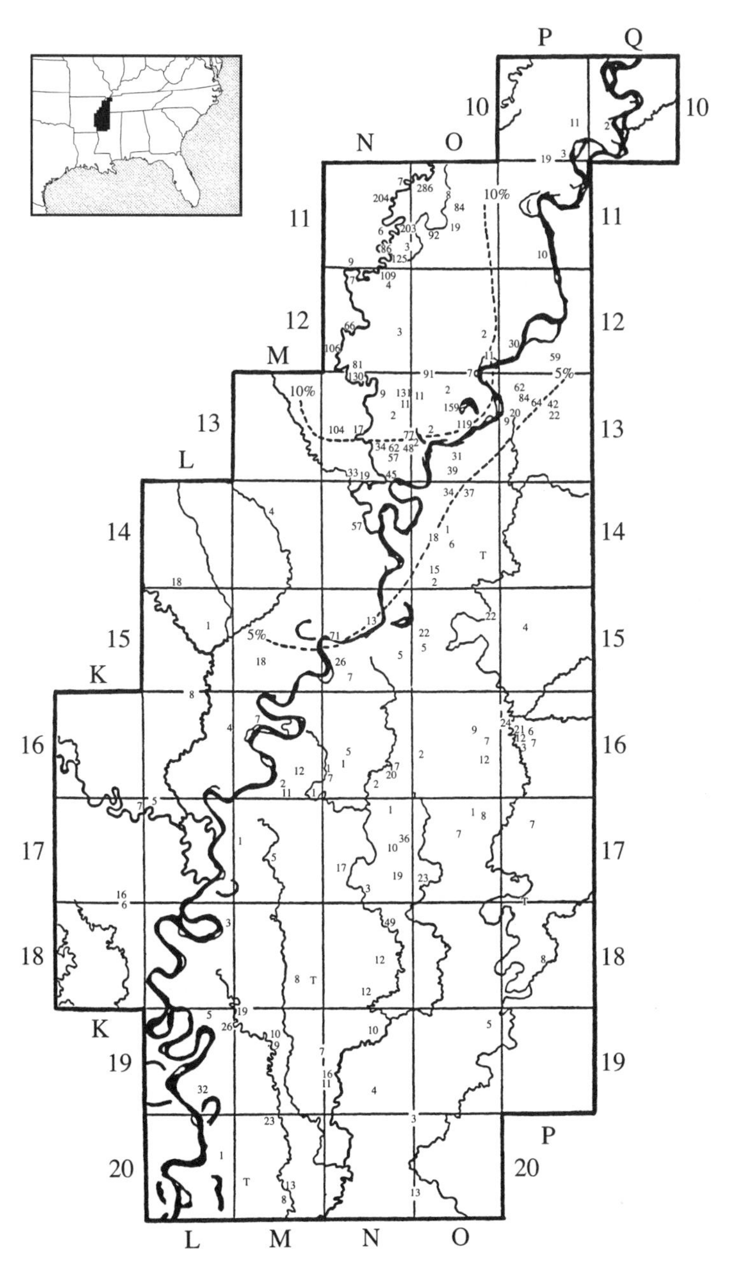

Figure 2.10. Percentage distribution of the type Parkin Punctated in the Lower Mississippi Alluvial Valley Survey area. Numbers refer to percentages at individual sites: 106, for example, means that the surface collection made at that site contained 10.6% Parkin Punctated sherds. Note that the percentages are highest in the northern portion of the region and taper off to the south (after Phillips *et al.*, 1951).

in the northern quarter of the survey area and fall off dramatically to the south. This indicated to the authors that the "heartland" of Parkin Punctated was around the lower reaches of the St. Francis River in northeastern Arkansas (the rows of squares labeled 11, 12, and 13 in Fig. 2.11). Similarly, Fig. 2.11 shows the distribution of Marksville Stamped, and the pattern is exactly the opposite; the percentages fall off sharply as one moves from south to north. In fact, the heartland of Marksville Stamped is east-central Louisiana (Phillips, 1970), which fell to the south of the region surveyed by Phillips, Ford, and Griffin.

In his slightly later monograph entitled "Measurements of Some Prehistoric Design Developments in the Southeastern States" Ford (1952) was explicit about the influence of the spatial dimension. There will be a time lag, he noted (Fig. 2.12), between the age of a type in its area of origination and its age in the area(s) to which it diffuses, a phenomenon later given the misnomer the "Doppler effect" by James Deetz and Edwin Dethlefsen (1965) (see also Dethlefsen and Deetz, 1966). Ford suggested that the direction of diffusion could be ascertained (1) by the decreasing relative frequency of a trait or type as one moved away from its area of origin or (2) more "reliably" by the detection of ancestral forms of the trait or type in the area of origin.

Phillips and his colleagues also noted another analytical problem caused by space, and it was an insidious one: As they moved farther away (in time and/or in space) from the "centers" that produced the archetypal pottery used by archaeologists to construct the type,

> the characters that we have selected as determinants for the type gradually shift, the all-too familiar phenomenon of "creep," until at some point we can stretch our original type definition no further and have to consider whether material "X" more closely resembles type "B," already established at another center, or whether it is not sufficiently like either "A" or "B" and must be given an independent status as type "C." These wretched hair-line decisions beset the classifier at every step. (Phillips *et al.*, 1951:65)

Phillips, Ford, and Griffin did not explore the spatial variation in their types, except to plot for some types the percentage occurrence of those types at sites in the survey area (Figs. 2.10 and 2.11). In a few instances they noted differences in surface treatment that might form the basis for future type subdivision. They believed that creation of additional types might be warranted, but the thought of such an occurrence left them pessimistic:

> The archaeologist who thinks he has achieved a final classification of anything is a rare and probably untrustworthy individual. Most of the shortcomings of our classification have been fully exposed in the type descriptions. Our guess is that very few of our types will stand up when more and better material is available. Many of them will break down into more specialized groups, a few (we may hope) will be combined into more general groups. It is not likely that the total number of types will be reduced. The outlook for the Southeast as a whole, so long as present typological methods remain in favor, is not pleasant to contemplate. Where we are now counting types in tens, they will be counted in hundreds. (Phillips *et al.*, 1951:426)

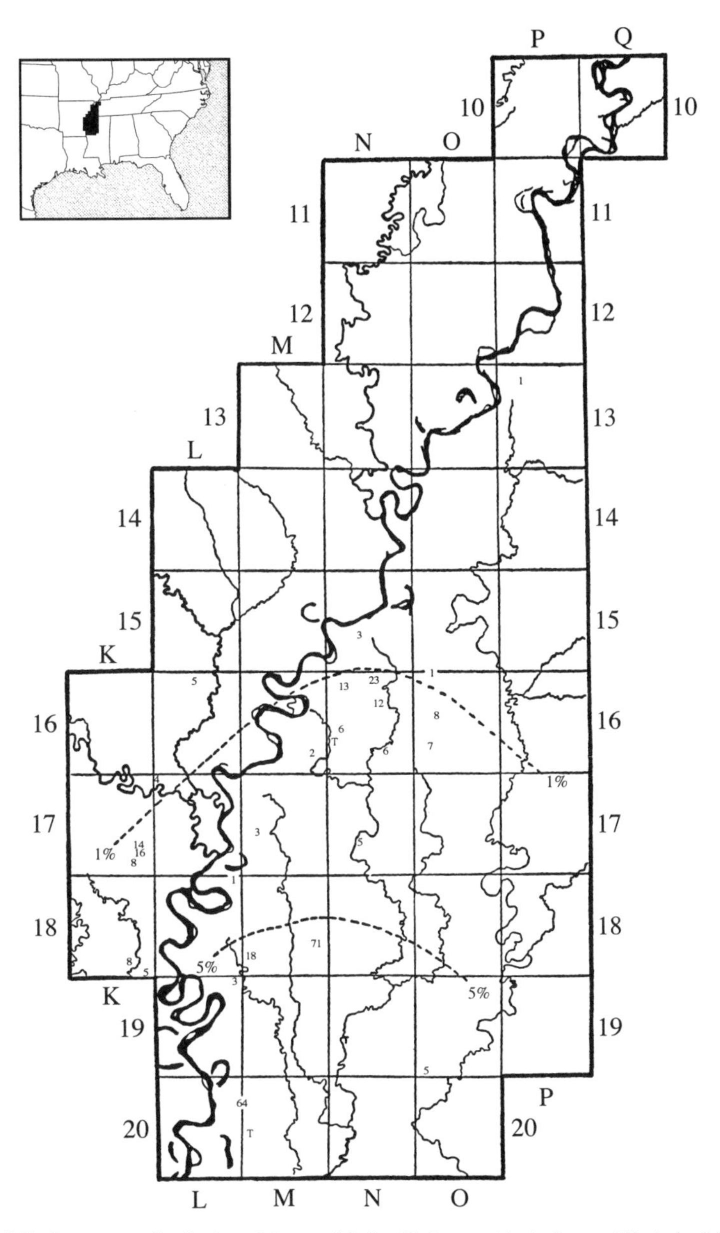

Figure 2.11. Percentage distribution of the type Marksville Stamped in the Lower Mississippi Alluvial Valley Survey area. Numbers refer to percentages at individual sites: 71, for example, means that the surface collection made at that site contained 7.1% Marksville Stamped sherds. Note that the percentages are highest in the southern portion of the region and taper off to the north (after Phillips *et al.*, 1951).

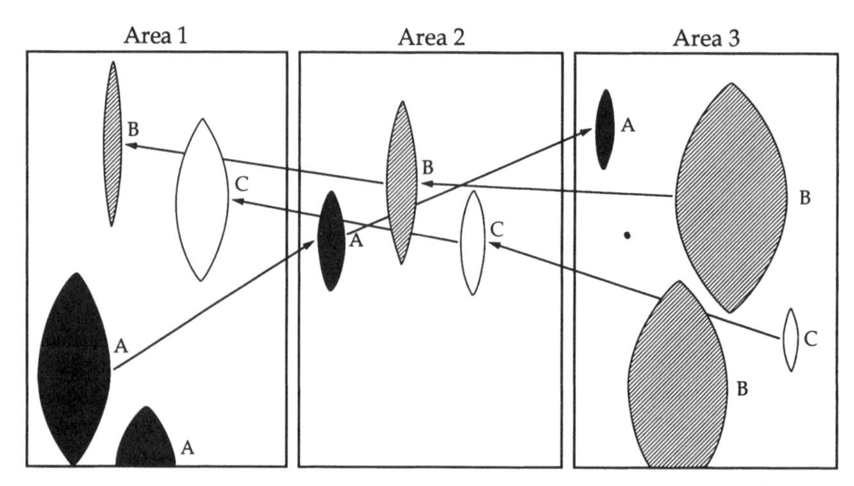

Figure 2.12. James A. Ford's model of the diffusion of types across space, illustrating change in frequency and time lag, phenomena Ford believed must be considered in making temporal alignments of relative chronologies. Each letter and kind of shading denotes a distinct type (after Ford, 1952).

Phillips (1970), in his monumental summary of the Mississippi River valley from the junction of the Mississippi and Ohio Rivers to the Gulf of Mexico, later tried to address the issue of spatial variation by expanding the list of types and by subdividing many of them into individual varieties. He took the original 20 ceramic types established during the earlier survey and which dated to the Mississippian period, expanded it to 40, and added 88 varieties. He created five varieties out of Parkin Punctated (Fig. 2.13) alone. For the earlier Marksville Stamped type, he created seven varieties (Fig. 2.14). Types, Phillips reasoned, should be formed primarily on the basis of three dimensions of variation: paste (e.g., clay-tempered, fine-shell-tempered), surface treatment (e.g., cord marked, plain surface), and design (e.g., punctated, incised). A secondary set of dimensions—Phillips's modes of form and modes of decoration—is used to create varieties. Phillips's view was that types are directly related to widespread regional expression of historical relations, that is, historically related peoples made and decorated their pottery similarly, and that varieties within specific types reflect areal and temporal variations in the norm of the types.

Phillips's assumption might not be unreasonable, though he offered no rationale for it (see Davis, 1981, for an early formulation of such a rationale and Lipo *et al.*, 1997, for a detailed theoretical rationale). Beyond that, there is the problem of deciding what someone means when he or she says that a particular sherd is such-and-such a variety of such-and-such a type. Often there is little consistency among investigators in terms of how they define particular types and

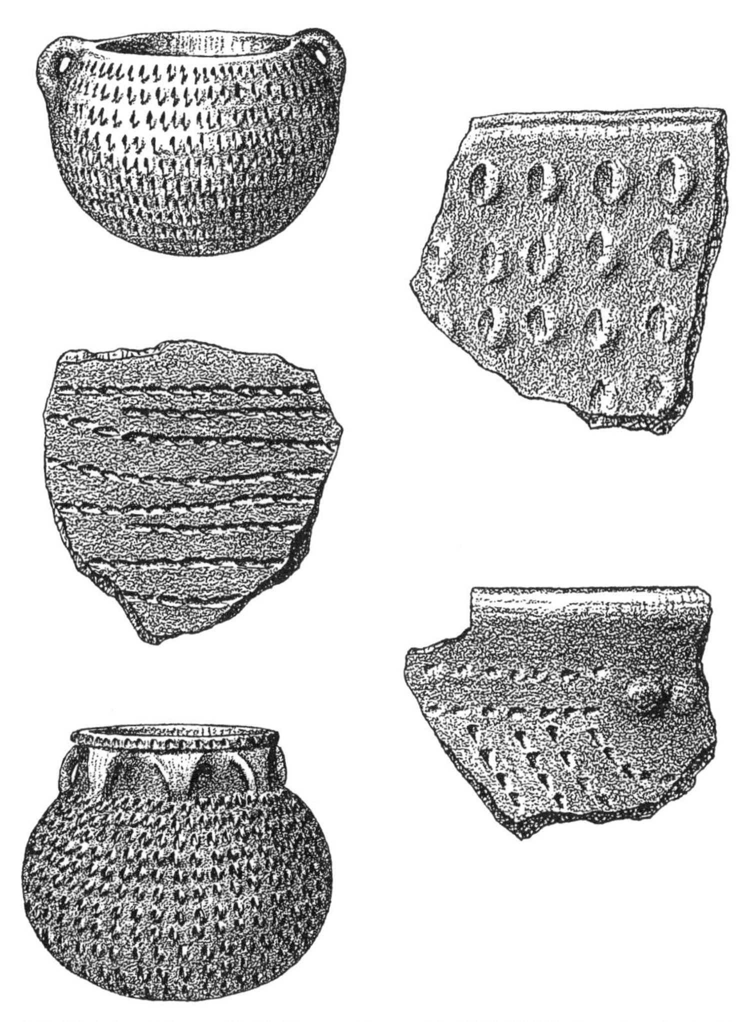

Figure 2.13. Varieties of the type Parkin Punctated created by Philip Phillips based on sherd collections from the central Mississippi River valley (after Phillips, 1970).

varieties. Archaeologists such as Phillips, Ford, and Griffin (1951) regularly mention type "descriptions" and do not specify how or if their descriptions differ from sets of definitive criteria for types. That is, describing a pottery type, for example, as comprising specimens that are "incised with parallel lines, shell-tempered, red-colored on the interior surface, and less than 1 cm thick," does not tell us which, if any, of these attribute states is necessary and sufficient for

Figure 2.14. Varieties of the type Marksville Stamped created by Philip Phillips based on sherd collections from the lower Mississippi River valley (after Phillips, 1970).

categorizing a new specimen as a member of this type. This problem, in conjunction with the ever-expanding number of types and varieties, tends to create a confusing situation.

Which Are Real: Types or the Objects Placed in Them?

Although we have not called attention to it, we have maintained a strict distinction throughout the discussion between type as an abstraction—an ideational unit—and the actual objects placed in a type. For example, we earlier noted that, "At present we cannot say for sure whether points of one type preceded those of another"; we did not say "whether one type preceded another." Logically, types cannot precede types, since types are nothing but abstractions. But the points placed in a type, given that they are empirical, can precede other points in terms of when they were made. As a shorthand notation, archaeologists often speak colloquially of type A preceding another type, but most are aware of the difference between theoretical and empirical units. This point was made beautifully by Phillips, Ford, and Griffin (1951:66):

> Exigencies of language require us to think and talk about pottery types as though they had some sort of independent existence. "This sherd *is* Baytown Plain." Upon sufficient repetition of this statement, the concept Baytown Plain takes on a massive solidity. The time comes when we are ready to fight for dear old Baytown. What we have to try to remember is that the statement really means something like this: "This sherd sufficiently resembles material which *for the time being* we have elected to call Baytown Plain." Frequent repetition of this and similar exorcisms we have found to be extremely salutary during the classificatory activities.

Unfortunately, most archaeologists since the time Phillips, Ford, and Griffin published their monograph have forgotten to repeat, if only to themselves, the "exorcism" quoted above. Types, which they (especially Ford) viewed as arbitrary constructions, have come to be treated as "real," empirical units that had been discovered as opposed to created. In other words, types are treated as if they are natural kinds, and thus have essences, which they clearly do not. Rather, they are ideational units used to measure empirical units. Ford realized this, but apparently Phillips (1970) did not.

The confusion over empirical and ideational units exists because there are two basic ways of viewing kinds, or types. One way is to view them as natural units that have specific essences that identify them as X, Y, or Z. Each essence is, in fact, so overwhelming that there can be no confusing one kind with another; any specimen with a particular essence is lumped in with all other similar specimens. In effect the essence of each individual specimen placed in that group is also the essence of the group as a whole. The other way of viewing kinds is to treat them as collections of things that hold properties in common. Under this view, for example, there is no essence that makes a Thebes point a Thebes point. Rather, points placed in that category simply share properties that we deem to be of importance, for whatever reason, for placing them in that category as opposed to another.

The issue of essences versus properties in common is an important one in any historical science. Biologists have long wrangled over the meaning of types, or kinds. Are species best viewed as natural units (e.g., Kitts, 1984; Schwartz, 1981) or as collections of individuals that look similar and that share the same isolating mechanisms that keep them from reproducing with other organisms (e.g., Ghiselin, 1981; Hull, 1978, 1980; Mayr, 1963, 1987; Sober, 1980, 1984)? Our opinion, shared by the majority of biologists, is that species are the latter. The confusion results from mixing essences with properties in common:

> To be sure every essence is characterized by properties in common, but a group sharing properties in common does not need to have an essence. The outstanding characteristic of an essence is its unchanging permanence. By contrast, properties in common of a biological group may be variable and have the propensity for evolutionary change. What is typical for a taxon may change through evolution at any time and then no longer be typical. (Mayr, 1987:155)

Although we do not go into detail here, the critical relevance of this argument to archaeology follows directly from the view that behaviors and products of behavior are parts of human phenotypes in the same way that birds' nests are phenotypic (O'Brien and Holland, 1990, 1992). How can any phenotypic feature have an essence, since it is constantly evolving as the organism (including humans) evolves? The answer is, it cannot. Attributes that show up in the archaeological record must be viewed instead as constantly changing entities that are as subject to evolutionary processes as is any somatic feature. Does this mean that archaeological kinds cannot be created? No, of course not. We can create kinds, but it cannot be overemphasized that these are strictly ideational units. Ford did not make reference to phenotypes in his classic treatment of types, but he knew that types were first and foremost analytical units. Phillips, on the other hand, viewed them as discoverable, and thus empirical units.

Until the early 1950s, most archaeologists were content to worry about the chronological placement of the types they arbitrarily netted from various cultural streams or lineages. It was thought that those units might reflect the cultural norms or customs of the people who made the artifacts being placed in the types, meaning that the units might in some way be "real," but such suspicions were merely commonsense rationalizations for the units. Types were not empirically testable except in a tautological manner, such as in George Brainerd's (1951a) notion that the popularity of cultural norms produced normal frequency distributions of types through time, a principle that extended back to the second decade of the twentieth century and the work of Kroeber, Nelson, and others. This principle became tautological when normal frequency distributions were used to "prove" the existence of norms or customs by Brainerd (1951a) and others. In the absence of any theory explaining why artifact types might act as some of them did, common sense prevailed, and the possibility that types were somehow real units grew stronger.

Ford's types had a somewhat arbitrary appearance, and thus might or might not reflect cultural or ethnic affiliation, even though they clearly were historical types that measured the passage of time. Albert Spaulding, on the other hand, no doubt encouraged by Walter Taylor's (1948) admonition that archaeology should be anthropological, wanted a typological system that "expressed at one stroke the classifier's opinion of the cultural relationship *and* the chronological position of an assemblage," as such a system would allow "a combined presentation of [the] independent units of chronological position and cultural affinity" (Spaulding, 1949:5). This was a lofty goal: the creation of a kind of unit that marked not only time but also ethnicity. Earlier archaeologists knew that many of their types measured time, and in some cases they assumed that those same types measured ethnicity, but they, like Spaulding, were unsure how best to construct such a dual purpose unit.

Spaulding became sure as a result of a paper Brainerd published in 1951. In it, Brainerd (1951b) made two observations that undoubtedly influenced Spaulding (Lyman and O'Brien, 1997; Lyman *et al.*, 1997b; O'Brien and Lyman, 1998). First, he indicated that "Archaeological taxonomy [read *typology*] is in itself a generalizing procedure which ultimately depends for its validity upon the archaeologist's success in isolating the effects of culturally conditioned behavior from the examination of human products" (Brainerd, 1951b:117). His second point was that

> The first step of procedure in artifact analysis is usually the formulation of types, groups of artifacts, each of which shows a combination of similar or identical attributes or traits.... If [the observation quoted above] is acceptable, the systematics used must have cultural validity in that they must mirror the culturally established requirements met by the artisans. In his search for these tenets of the unknown group it behooves the archaeologist as a scientist to work objectively, free of a priori conceptions. *The attributes used in sorting artifacts into types should thus be objectively chosen as those which occur most often in combination in single artifacts.* Criteria based upon subdivisions of an attribute which occurs in a continuous range through the material are preferably used only when the distribution curve of the attribute in the archaeological samples shows binodality, and *the dividing line for sorting should be drawn between the nodes.* By use of the above requirements for type attributes, the archaeologist can *objectively describe the cultural specifications followed by the artisans.* Statistical procedures for the formulation of, and sorting of specimens into, types satisfying these requirements are feasible, and may in some cases be useful. It seems conceivable also that mathematical studies of attribute combinations may demonstrate more finely cut cultural differentiation without the use of the intermediate concept of types, for types are, after all, simplifications to allow qualitative division of the material into few enough categories to permit inspectional techniques of analysis. (Brainerd, 1951b:118–119; emphasis added)

Brainerd (1951b:124) argued that "it is conceivable that a bridge may be found uniting the objectivity of the taxonomist to the cultural sensitivity of the humanist.

Cultural intangibles can, if they exist, be made tangible," and he concluded that "[b]etter technique is the solution" to archaeology's dilemma. There is, of course, nothing wrong with being simultaneously objective and culturally sensitive, as long as we realize that the latter has absolutely nothing to do with science.

Spaulding (1953) used statistics to launch his attack against the perceived arbitrariness of artifact types, attempting to demonstrate that types formed by clustering algorithms had sociobehavioral meaning. His definition of a type as "a group of artifacts exhibiting a consistent assemblage of attributes whose combined properties give a characteristic pattern" (Spaulding, 1953:305) was compatible with earlier definitions, such as that by Irving Rouse (1939) and even that by Ford (1938b) because of its emphasis on the recurrence of attribute combinations. The difference, however, was that for Spaulding, recurrence was empirically, and "objectively" through statistical analysis, determined on a closed set of materials. Spaulding worked with sherds from one site at a time, measuring similarities between sherds in terms of attribute combinations and then creating groups based on statistically significant recurrent patterning. Were the types represented at locations outside of the one that produced the sample that was analyzed? This question was impossible to answer because of the highly idiosyncratic, sample-specific nature of the types he created. If even a single sherd from outside the original sample were added to it, the entire exercise would have to be repeated.

Ford's types were also formed from recurrent combinations of attributes, but the pottery samples were much more widely distributed spatially (O'Brien and Lyman, 1998). Thus the types had temporal and spatial distributions. They were formed inductively by trial and error, but they were theoretical units nonetheless. Once formed, their analytical utility was tested deductively to determine whether they measured time; in Krieger's (1944) words, did they pass the historical significance test? Spaulding believed that his types were real and inherent in the specimens; in other words, they were empirical units. To Spaulding (1953:305), "Classification into types is a process of discovery of combinations of attributes favored by the makers of the artifacts, not an arbitrary procedure of the classifier." Because types are inherent in the data, they must be discovered inductively, and statistical techniques, as suggested by Brainerd, provide the objective means of determining which attributes regularly, in particular, more often than random chance would allow, co-occur on specimen after specimen. Since artifacts are products of human behaviors, discovery of recurring attribute combinations (types) is simultaneously discovery of that behavior. Spaulding's types could do no analytical work, nor could their reality be tested except by applying a different clustering algorithm, resulting in numerous arguments over the appropriateness of different algorithms (see the review in Dunnell, 1986a) and the accomplishment of little archaeological work.

In a comment on Spaulding's (1953) paper, Ford (1954a) protested that

Spaulding's approach was naive because it only suggested cultural norms; it did not help write culture history. In his response, Spaulding (1954:393) noted that Ford had not "challenged the validity of the techniques [Spaulding] used to discover [attribute] clusters" or types and underscored the procedural murkiness in Ford's constructions of "attribute combinations." Spaulding failed to see the difference between Ford's types as theoretical (definitional) units and his own types as idiosyncratic (empirical) groups. The latter were obviously different creatures from Ford's. Spaulding noted that his attribute clusters included inferences as to the behavior of the artifact makers, whereas Ford's did not. Such combinations existed as human creations and were sortable into recognizable, empirical sets; thus, they had to be real.

Ford (1954b) replied, and although a careful reading of his paper makes clear the distinction between Ford's and Spaulding's positions, it also points out the manner in which Ford conflated elements of the two perspectives. Cultural types certainly once existed, Ford thought, but he was not particularly interested in discovering them. Rather, he wanted "type groupings consciously selected [by the archaeologist to produce] a workable typology ... designed for the reconstruction of culture history in time and space" (Ford, 1954b:52). However, in his reply he did not specify how such groupings were to be extracted from the flowing, constantly changing cultural stream (see Fig. 3.7 and associated discussion), nor did he clarify that his units were ideational, whereas Spaulding's were empirical. Clearly, ideational units were called for that allowed one to measure culture change, and thus in lieu of theoretically informed unit construction, trial and error was used. This was, in fact, how archaeological types had almost always been formed. To Ford, significant formal variation existed at any point in the time–space continuum, and although that variation might "tend to cluster about a mean, which [the analyst] could visualize as the central theme of the type ... [he or she] cannot rely upon the culture bearers to define this theme. They may or may not be aware of it.... The [type], then, is an abstraction made by the [analyst] and derived from cultural activity" (Ford, 1954b:45).

In Ford's view, discontinuities along either the temporal or spatial dimensions of the archaeological record presented the archaeologist with natural seams, at which points the time–space continuum could be broken up into chunks. Ford was proposing that discontinuities presented convenient points at which to insert arbitrary breaks in the continuum; if such discontinuities did not present themselves, one could make the cuts at any arbitrary points. Ford's critics, especially Phillips (e.g., 1970), never understood how Ford could make temporal breaks when no natural divisions, either stratigraphic disruptions or the appearance of new cultural traits, presented themselves. Ford could do it because in his mind the flow of time, and hence of culture, was seamless, punctuated only by changes in tempo (O'Brien and Lyman, 1998). Chunks had to be carved out of the continuum for analytical purposes, but there was nothing particularly real about them. Types, to Ford, were nothing but accidents of the samples available: "[T]he particular locality where an archeological collection chances to be made will be one of the factors that determines the mean and range of variation that are demonstrated in any particular tradition in the culture that is being studied" (Ford, 1954b:49). Chance samples of the continuum would provide discontinuous snapshots of that continuum; hence, types "are easily separable and they look natural [that is, 'real']" (Ford, 1954b:52).

In the end, Ford's strategy for refuting Spaulding's position fizzled because his allusions to customs and standards gave Spaulding's types a certain reality. Despite Ford's poorly constructed arguments and obtuseness, his basic position stands in stark contrast to that of Spaulding, whose types were "more or less discrete packages" (Dunnell, 1986a:181). Any variation not assignable to such packages lacked explanatory significance. Comparison of Spaulding's "real" types must be qualitative and must focus on differences between them. Ford's types stem from a conception of reality that views variation in form as being continuous across space and through time. Division of that continuity into chunks through the use of ideational measurement units—types—is a trial-and-error process, the successes of which Ford chose to evaluate, we believe correctly, with the historical significance test (Krieger, 1944).

The Ford–Spaulding debate did little to clear up the confusion over the nature of types, and in fact some archaeologists (e.g., Cowgill, 1963) saw little or no contradiction in the two positions. Given the general belief that types probably have some cultural meaning, it is not surprising that Americanist archaeologists took typology a step further and broke types down into varieties in an attempt to get at smaller-scale sociological phenomena. James Gifford (1960) suggested, on the one hand, that a variety represents individual or small-group social variation within a society, a position taken to its extreme in Phillips's (1970) treatment of pottery from the lower Mississippi River valley. On the other hand, a type is

> the material outcome of a set of fundamental attributes that coalesced, consciously or unconsciously, as a ceramic idea or "esthetic ideal"—the boundaries of which were imposed through the value system operative in the society by virtue of individual interaction on a societal level. These ceramic ideas occurred in the brains of the potters who made the ceramic fabric that constitutes a type, and they are not by any means creations of an analyst. (Gifford 1960:343)

Gifford's rationalization of his units was the typical one: Potters "tend to conform to the demands of a majority of the norms that are a part of their culture at a particular time in history" (Gifford, 1960:343), and cultural phenomena are not randomly distributed in time and space (Gifford, 1960:342). Gifford (1960:342), like Spaulding before him, believed that an inherent order in the data was discoverable and that "Classificatory schemes ... are in part useful as a means toward this end." This was a patently essentialist outlook.

TYPOLOGY IN RETROSPECT

The culture historical period in Americanist archaeology, from roughly 1910 to 1960, was, as we have seen, the heyday of typology. To be sure, historical types are still being created today as we try to bring some semblance of chronological order to the archaeological record, but in most respects the groundwork for how such types are formed was laid decades ago. The eventual standardization of terms that was missing prior to about 1915 may have eased communication between and among archaeologists, but it had another, lasting consequence. For Ford and the host of archaeologists who followed him or were his contemporaries, historical types were formulated on the principle of "use whatever works." Unfortunately, although Ford and his colleagues had labored to establish terms that could be used consistently, the results were anything but consistent from region to region and from investigator to investigator. It is important to point out why this was and continues to be the case.

Robert Dunnell (1986a) produced perhaps the best exposition of the problem. He points out the dimensional (see Dunnell, 1971) nature of early twentieth-century Americanist artifact classifications, particularly pottery classifications, wherein each dimension is an analytical unit and within each dimension is a set of nonredundant attributes:

> Although the list of dimensions varied somewhat from area to area based on tradition and the variable character of archaeological materials, this approach provided that a type hold some value in each of a standard set of dimensions of observation and measurement. Just as importantly, beyond setting limits on the dimensions that could be used, this approach did not provide any guidance in the *selection* of definitive attributes. They were literally *descriptions of pottery assigned to a particular type*, not definitions of the type. While a value or set of values was required for each dimension (e.g., a particular paste texture, a particular temper, or a particular exterior and surface treatment), which dimensions were definitive varied from type to type and were unmarked in the descriptions themselves. Thus, type descriptions embodied a very substantial amount of operational ambiguity. One had to know, case by case and a priori, which attributes were cause for the assignment of type and which were simply attributes associated typically with the definitive elements. Because the type descriptions were in fact descriptions of particular pottery assemblages (how else could one obtain a "thickness" range, for example), a further and even more far-reaching structural ambiguity was introduced. Were types the creations of archaeologists and, thus, definitionally associated sets of attributes (i.e., analytic tools for the dissection of the archaeological record), or were they empirically associated sets of attributes that "discovered" existential entities? Either interpretation was possible, and both would be pursued. (Dunnell 1986a:165–166)

Both interpretations *were* pursued—the types as "creations of archaeologists" by Ford, and the types as "'discovered' existential entities" by Brainerd and Spaulding—and still are. Further, many archaeological types, whether created or discovered decades ago or last year, are idiosyncratic units based on a few samples from a few sites, while others, although based on large samples from numerous sites, tend to be catchall categories into which can be thrown everything except the kitchen sink. Again, it is beyond the scope of this book to consider all the intricacies of archaeological systematics. Many papers and several books are available that cover the topic in detail (e.g., Adams, 1988; Adams and Adams, 1991; Dunnell, 1971, 1986a; Klejn, 1982; Whallon and Brown, 1982). In our view, J. O. Brew's (1946:46) early remarks that (1) classifications are arbitrary constructs of the analyst and "no typological system is actually inherent in the material"; (2) there is no ideal all-purpose classification system that allows the solution of a set of heterogeneous problems; and (3) the classification system used should fit the purpose of the investigation (see also Steward, 1954) still hold over half a century later. In short, types should be constructed with a particular analytical problem in mind. In the context of this book, the analytical problem is to measure time, and thus historical, or chronological, types are what need to be constructed.

Many of the archaeological types in use today, regardless of the shaky grounds on which they were formed, are good chronological markers. In short, they pass the historical significance test; each comprises specimens made during a single relatively short interval of time, and the frequency distribution through time of the specimens comprising a historical type approximates a unimodal curve. Much of the credit for this success can be laid at the feet of Nelson, Griffin, Ford, and their contemporaries, all of whom were smart enough to use whatever attribute combinations—types—worked in bringing chronological order to the archaeological record. They laid the groundwork—Nelson in the Southwest and Ford and Griffin in the Midwest and Southeast—that everyone else could follow. They developed a purely archaeological device—variation in artifact form rendered as historical types, or what became known as "styles"—that monitored the passage of time. How they knew that device measured time is a critical part of our disciplinary history. It also is a complex part of our history that is sorted topically only with difficulty. The order of topics (methods of relative dating) we follow in the next four chapters should not be taken as indicative of our or the discipline's preference or of some inherent order of ascending or descending significance. Rather, it is merely a convenient order. Central to all four chapters is the role of chronological types, the most basic and truly the only purely archaeological tools used to measure the passage of time.

3

Seriation I

Historical Continuity, Heritable Continuity, and Phyletic Seriation

Over two decades ago, William Marquardt (1978:257) observed that, "Seriation is a deceptively simple technique, the unnecessarily abstruse discussion of which has become an embarrassment to quantitative archaeologists. At the same time, seriation is a mathematically entertaining and eminently useful procedure, the epistemological status of which has been insufficiently explored." We agree with Marquardt: Seriation as a method of ordering phenomena based on their similarity is deceptively simple, and usual archaeological practice has been to make it as abstruse as possible. Recall from Chapter 1 that within the numerous books on archaeological dating methods, very few pages are devoted to seriation. It rarely is taught anymore except in the most perfunctory manner, with minimal effort to explore either its roots in Americanist archaeology or its interesting epistemological underpinnings.

More than any other method we can think of, seriation has had a complex history. This can be attributed partly to shifts in attitude over the utility of seriation as a chronological tool relative to that of superposed deposits and partly to the failure to keep straight the several different techniques of seriation. Textbooks usually credit British archaeologists for introducing seriation into the United States, but in reality only what we term *phyletic seriation* was in use in nineteenth-century Britain, and although it eventually was introduced to American archaeologists, it differs significantly from the kind of seriation invented by Americanists in the second decade of the twentieth century. That distinctly American kind of seriation is known as *frequency seriation*, which eventually led to the Americanist development of what is known as *occurrence seriation.*

In this chapter we outline the historical role seriation played in Americanist archaeology and attempt to clear up several misconceptions about what seriation is and is not. Textbook treatments not only conflate distinct seriation techniques, but also tend to confuse seriation with chronological observations based on excavation, making it sound as if vertical sequences of artifacts can be seriated. This is incorrect, but we cannot blame modern authors for perpetuating a confusion that originated with their predecessors. Conflation of different seriation techniques has had several consequences, not the least of which is an inaccurate reporting of the history of the method and the techniques that it comprises. More importantly, such conflation has caused the discipline to overlook a key distinction in how archaeologists measure time: continuously or discontinuously.

We also explicate the key assumptions underpinning the seriation method. As a method, seriation comprises a particular set of assumptions and their corollaries constructed to solve a problem, in this case, chronological ordering of artifacts. Particular applications of the method can take several different forms, or what we term techniques, yet each rests on the assumptions of the method. The most important assumption of the method is what we term *heritable continuity*, which is in turn assumed to be reflected by *historical continuity* rendered as the similarity of archaeological materials (O'Brien and Lyman, n.d.). We elaborate on these assumptions later. It suffices here to note that the different seriation techniques are distinctive because they are operationalized by measuring similarity, or historical continuity, and assuming heritable continuity in distinctive ways. To make the distinction clear, we begin with a discussion of what seriation is and then turn to a history of the method. We conclude this chapter with a discussion of the first seriation technique developed by archaeologists—phyletic seriation—and reserve discussion of occurrence seriation and frequency seriation for Chapter 4.

WHAT IS SERIATION?

Marquardt (1978:258) defined seriation as "a descriptive analytic technique, the purpose of which is to arrange comparable units in a single dimension (that is, along a line) such that the position of each unit reflects its similarity to other units." Seriation is a descriptive method that simply orders things in a row or column. Units used in a seriation must be comparable, which makes sense if we are searching for an order based on similarity in such things as artifact form or decoration. Marquardt's definition tells us nothing about the kinds of units that can be used in seriation, nor should it. As we will show, many kinds of ideational units can be used successfully. The trick is in understanding when to use which kind of unit, and for that we need to understand the basic differences among the various seriation techniques. Third, nowhere in Marquardt's definition is the term *time* mentioned. Seriation creates a linear order, but that order only tells us that the odds are good that two adjacent things are more alike than either is to things farther up or down the line. Seriation most definitely is used in archaeology to measure the passage of time, but whether or not the order of units created by a seriation is a sequence in the sense that it actually reflects the passage of time is an inference; it is not axiomatic. Culture historians working earlier this century understood the inferential nature of chronology based on a seriated sequence; witness A. L. Kroeber's (1916a:20) statement that the proof of the chronological significance of his seriated potsherds was "in the spade."

Seriation as a means of inferring chronology routinely is confused with other methods of chronological ordering. Part of the problem resides in the fact that similar graphing techniques were used in Americanist archaeology to show chronology obtained both by stratigraphic means and by seriation (Lyman *et al.*, 1998). Through time, the two methods became conflated, leading to the presumption that anytime artifacts or collections of artifacts were ordered chronologically, seriation was involved. For example, seriation has been defined as "the determination of the chronological sequence of styles, types, or assemblages of types (cultures) by any method or combination of methods. Stratigraphy may be employed, or the materials may be from surface sites" (Hester *et al.*, 1975:272). This is incorrect. Seriation is allied with stratigraphy in that both have long been used by archaeologists in developing chronological ordering, but we do not agree that "stratigraphy may be employed" in a seriation. For one thing, seriation measures time as a continuum, whereas stratigraphy measures it discontinuously (see Chapter 5).

John Rowe (1961:326) excluded the use of superposed strata in his definition of seriation: "[T]he arrangement of archaeological materials in a presumed chronological order on the basis of some logical principle other than superposition.... The logical order on which the seriation is based is found in the combinations of features of style or inventory which characterize the units, rather than in the external relationships of the units themselves." We prefer this definition precisely because it underscores that the ordering is based on formal attributes of the materials being seriated. That is, seriation is based on intrinsic properties or attributes of the artifacts and not on their relative vertical positions in a column of sediments; the last is an extrinsic property or attribute. Another way to say this is that seriation is a direct dating method, whereas superposition is an indirect dating method.

Preferring Rowe's definition, we cannot agree with statements such as that James A. Ford used seriation in his groundbreaking work in the Southeast during the 1930s (Trigger, 1989:200–202; Watson, 1990:43; Willey and Sabloff, 1993: 113–114); Ford rarely used the method (O'Brien and Lyman, 1998). Likewise, it often is claimed that Nels Nelson "for the first time made a strict use of statistical methods developed in Europe, and reported this method in 1916" (McGregor, 1965:42) and that this "statistical method" of studying fluctuating frequencies of types was introduced to the Americas as a result of the influence of W. M. F. Petrie (Browman and Givens, 1996:83). This is not true. Part of the confusion rests with the early culture historians who developed the methods; they themselves often failed to define and differentiate among various methods and techniques for measuring time. Sloppy use of terms resulted in misunderstanding the history of seriation, the various techniques by which it can be implemented, and its relation to the use of artifacts contained in superposed sediments for chronological purposes.

What is important to realize is that between about 1915 and 1935, several different terms were being applied to the same analytical technique in Americanist archaeology; simultaneously, the same or a similar term was being applied to distinct techniques. Early seriations were completely unrelated to such things as vertical or stratigraphic position in a site. Any chronological implications derived from seriation were implications of continuous time, but once stratigraphy and superposition became strategies for creating chronologies of artifacts as opposed to strategies for confirming that an ordering of artifacts in fact measured time's flow, archaeologists were measuring time discontinuously.

SERIATION IN AMERICANIST ARCHAEOLOGY

As far as we have been able to discover, anthropologist Edward Sapir (1916: 13) was the first Americanist to use the term seriation, indicating that "cultural seriation" was a "method ... often used to reconstruct historical sequences from the purely descriptive material of cultural anthropology." Importantly, he also stated that (1) the "tacit assumption involved in this method is that human development has normally proceeded from the simple or unelaborated to the complex"; (2) "evidence derived from seriation ... fits far better with the evolutionary than with the strictly historical method of interpreting culture"; and (3) this method "is probably at its best in the construction of culture sequences of simple-to-complex type in the domain of the history of artifacts and industrial processes, particularly where the constructions are confined to a single tribe or to a geographically restricted area" (Sapir, 1916:13–15). Cultural seriation was founded in the cultural evolution of such social scientists as Lewis Henry Morgan (1877) and Edward B. Tylor (1871): Primitive meant older, and advanced meant more recent. Some early archaeological seriation was based on this presumed course of cultural change, such as A. V. Kidder's (1915) suspicion that glazed pottery was more recent than unglazed pottery because of the greater technological sophistication of the former. Missing from such formulations was any discussion of how the assignments of primitive and advanced were to be made, these being such commonsense notions that discussion was unnecessary.

Leslie Spier (1917a,b) apparently was the first American archaeologist to use the term seriation, but without reference to Sapir's paper. Spier used the term to refer to Kidder's (1915) work in the Southwest, characterizing it as "the hypothetical seriation of several pottery techniques" (Spier, 1917a:252). Spier also used the term to refer to the work of his mentor, Kroeber (1916a,b), in the Southwest, though he characterized that work as "the hypothetical ranking of surface finds and the observation of concurrent variations [in the frequencies of types]" (Spier, 1917a: 252). We take up Kroeber's observations in Chapter 4; here the important point is that Spier characterized the work of Kroeber and Kidder differently. In fact, Kidder's (1915, 1917) seriations were of a decidedly different sort than Kroeber's. Spier's (1917a:252) crediting of Kidder for the "concept of seriation [and] Kroeber for ranking and concurrent variation" should have precluded any later confusion of two distinct analytical techniques, but this was not the case.

Although recognized for his use of what became known as frequency seriation when awarded the Viking Fund Medal in anthropology, Spier, like his contemporaries, did not explicitly define "seriation" in his seminal papers (Spier, 1917a,b). He later characterized seriation as a method in which the "remains of a stylistic variable (such as pottery) occurring in varying proportions in a series of sites are ranged [ordered], by some auxiliary suggestion, according to the seriation [ordering] of one element (one pottery type)" (Spier, 1931:283). Although this was in fact what Kroeber (1916a,b) had done, others who later used frequency seriation ordered their collections on the basis of multiple types. The "auxiliary suggestion" to which Spier (1931) referred, which earlier was characterized by him as a "principle for the seriation of the data" (Spier, 1917a:281), comprised the expectation that the relative frequencies of pottery types through time would exhibit smooth changes that approximated a normal distribution (Fig. 2.2). This suggestion is the "popularity principle" (Lyman *et al.*, 1997b:43) and to this day has served as the underlying guide—the axiom—to performing a frequency seriation, that is, ordering collections of artifacts using relative frequencies, or proportions, of artifact types.

The creation of terminological confusion cannot be laid solely at Spier's feet. Kidder (1919:298) characterized Spier's (1917a, 1918a, 1919) work as involving (1) the "seriation" of artifact collections on the basis of a single type of artifact and (2) subsequent testing of the validity of the final arrangement on the basis of "concurrent variations in [the frequencies of] the accompanying wares." Analytically, Spier (1917a,b) was simply mimicking what Kroeber (1916a,b) had done: ordering collections based on frequencies of types. Yet Kroeber did not originally refer to his particular analytical technique as seriation. Kroeber (1925:406) later referred to some of his own seriations as "non-stratigraphical comparison of the frequency of several types of ceramic decoration"; these are correctly categorized as frequency seriations.

Willey and Sabloff (1993:113) suggest that Kroeber popularized and "made explicit" the notion of what we refer to as phyletic seriation in a series of papers written in the 1920s in which he and two of his students, William Duncan Strong and Anna Gayton (e.g., Gayton, 1927; Gayton and Kroeber, 1927; Kroeber and Strong, 1924a,b; Strong, 1925), analyzed Max Uhle's collections of Peruvian pottery. Kroeber and his students used phyletic seriation to order Uhle's collections chronologically, but there was precedence for such in Kidder's (1915, 1917) work as well as in anthropology generally (e.g., Sapir, 1916; Wissler, 1916c). Kroeber (1927:626) used the term "stylistic seriation" to refer to Uhle's (1902, 1903) earlier analysis of the Peruvian material. Uhle (1902:754) did not use the term seriation, referring to what he did as the "method applied by Flinders Petrie in Egypt to prove the succession of styles by gradually changing character of the contents of graves."

Petrie (e.g., 1899a,b, 1901) termed what he did "sequence dating" (see below), a term repeated by few Americanists (e.g., Heizer, 1959; Hole and Heizer, 1973). Praetzellis (1993:76) states that "Seriation was developed by Flinders Petrie for the analysis of excavated Egyptian ceramics, and apparently brought to North America by Max Uhle who introduced it to Alfred Kroeber." This is a common misconception in Americanist archaeology (e.g., Browman and Givens, 1996); Trigger (1989:202) is correct when he notes that "although Kroeber may have learned the basic principles of typology and seriation from [Franz] Boas and known of Petrie's work, his technique of seriation was not based on the same principles as Petrie's."

Given the discussions of Kidder, Kroeber, Spier, and others, it is not surprising that history is confusing. To help cut through the confusion, Fig. 3.1 presents a taxonomy of seriational techniques. Following Rowe, we first divide the method into two major kinds: *similiary seriation*, or "seriation by resemblance," and *evolutionary seriation*, or seriation by "a rule of cultural or stylistic development" (Rowe, 1961:326). We then subdivide similiary seriation into frequency seriation, occurrence seriation, and phyletic seriation. Similiary seriation is ordering based on no rules of development, whereas evolutionary seriation is based on such rules. Rouse (1967) proposed the term "developmental seriation" to avoid the connotation of evolution according to a rule, but we do not find the term helpful in this respect. We prefer the term "phyletic seriation," which implies a lineage, a chronological line of suspected heritable continuity rendered as similarity, but no rule of developmental direction. We might have chosen the term "phylogenetic

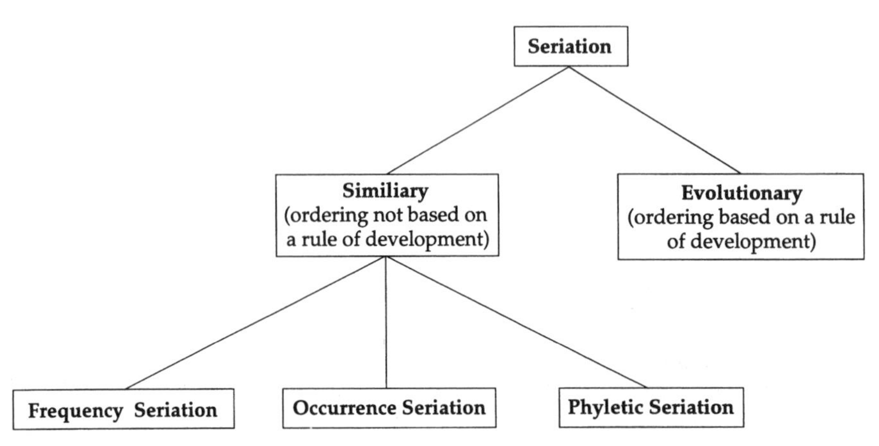

Figure 3.1. A taxonomy of seriation techniques. Seriation comprises techniques of ordering based on formal similarities (after Lyman *et al.*, 1998).

seriation" instead of phyletic seriation, but there are two very good reasons for not choosing the former. First, phylogenetic denotes branching, whereas phyletic merely denotes a line of descent, or lineage (Mayr, 1959). Second, phylogenetic connotes genetic descent. As we will see later, archaeologists have expressed displeasure with any genetic connotation in archaeological terminology. Our use of phyletic, then, is a compromise, and we emphasize that the term can entail linear, branching, or reticulate descent.

Evolutionary seriation is, technically speaking, a kind of similiary seriation because it, too, orders phenomena based on their formal similarities. The more alike two phenomena are, the closer they are placed to one another in the ordering; the less alike they are, the more distant from one another they are placed in the ordering. The singular significant difference between evolutionary seriation and the techniques we have placed under the term "similiary seriation" in Fig. 3.1 is that the former orders phenomena according to a rule of development, whereas the other techniques presume no such rule or direction. In other words, evolutionary seriation presumes a particular direction to the flight of time's arrow, whereas the other techniques presume nothing about that direction. Typical rules of development that have been used in evolutionary seriations include increases in technological complexity through time and artistic shifts in motif from naturalistic to stylized designs. Because such rules are now known to be historically contingent, and thus case specific, they are not often used as general principles of ordering, and we do not consider evolutionary seriation further.

THE KEY ASSUMPTIONS: HISTORICAL AND HERITABLE CONTINUITY

The seriation method involves placing objects or sets thereof in an order based on their formal similarities. The more attributes two phenomena share, the closer they are to one another in the order; the fewer they share, the more distant from one another they are in the order. The implicit but key assumption to the procedure, regardless of how similarity is measured, is that propinquity in formal properties denotes propinquity in time. This is the assumption of *historical continuity*. One might well ask why formal similarity should denote temporal similarity. In short, the assumption of historical continuity rests on the assumption that formal similarity is the result of heritable continuity. To borrow Darwin's (1859) famous phrase, seriation as a method for measuring time presumes "descent with modification." Similarity of descendant phenomena to ancestral phenomena is the result of a geneticlike connection between the two: Like begets like. As we will see, things are hardly this simple, but this is the crux of the method.

Given its significance, the preceding warrants reiteration. The seriation method as a chronological tool rests on the assumption that things produced at any particular moment resemble in significant respects things of the same kind produced at slightly earlier moments, just as they will closely resemble, in different ways, those produced at slightly later moments. For example, the model of phyletic seriation holds that objects of kind A produced at moment 3 will most resemble those produced at moments 2 and 4 and will resemble less those things produced at moments 1 and 5. Given this, let us say we have an array of objects of a specific kind, all of which were produced over some span of time. Further, let us assume that each sequential object produced resembles its predecessor, as it will its successor, in numerous ways. The problem is that the set of objects is jumbled, and our job is to put the objects in order based on their formal similarity. The principle of ordering dictates the procedure: place the objects most similar to one another adjacent to one another; as similarity between objects decreases, increase the distance between them. The inference that the order is chronological is founded on the principle of historical continuity.

Whether or not the order we produce is in fact a chronological ordering of forms—a temporal sequence—and thus measures time is an inference that must be tested with independent data such as finding the same ordering of forms in a sequence of superposed sediments. If the ordering is in fact chronological, and thus denotes historical continuity, then the next problem is to determine if the sequence is also a lineage, a line of heritable continuity. The two—heritable continuity and historical continuity—are not necessarily one and the same. If heritable continuity is indicated, and we discuss how this is determined both below and in Chapter 4, then historical continuity follows automatically. But if historical continuity, a mere temporal sequence of forms, is indicated, heritable continuity is possible and perhaps even highly probable, but it is not assured. In other words, historical continuity is founded in the similarities of the ordered phenomena; why the similarity exists may be due to heritable continuity or some other reason (we identify a significant other reason below). The two are separate inferences, though the latter underpins the ordering in the first place. Failure to distinguish and keep separate the two kinds of continuity conceptually and analytically has led to no end of problems in archaeology.

The assumptions underpinning seriation—historical continuity and heritable continuity—should sound familiar to anyone with a background in biology or paleontology, as they are the backbone of those disciplines. Biologists and paleontologists are vitally interested in historical and heritable continuity because they are central to understanding organismic evolution. But we do not need a background in evolution to understand these concepts and their value as chronological tools. Most of us, for example, would not have much trouble correctly ordering examples of a particular line of automobile. Experience has taught us that the year-to-year change in car design usually is not drastic. Manufacturers talk about "total redesign," but what they are really saying is that they made the new model a little shorter or a little wider and maybe gave the car more of an aerodynamic look. If ten people were shown 20 cars of the same line produced between 1978 and 1998, most of them could put almost all the cars in correct chronological order, regardless of how much or how little they knew about automobiles. Theoretically, they might not know which end of the sequence was the most recent, but they would get the basic sequence correct. The key to successful ordering would come from noting features that were exactly the same or nearly the same on different examples and putting those examples next to or close to each other in the sequence. This is simply the principle of heritable continuity in action.

We can do exactly the same thing with biological and archaeological specimens, regardless of how old they are. In biology this procedure is referred to as creating a phylogenetic history, and it has its roots deep in social science and natural history. A phylogenetic history is a seriation of forms of what are thought to be genetically related organisms. Various seriation techniques have seen increased use in the last decade by archaeologists who derive their theory from Darwinian evolution and seek to develop and explain historical lineages of artifacts. That group of archaeologists, of which we are a part, views historical analysis as fundamental to the study of evolution through the use of the archaeological record. Detailed analysis of artifact change also provides us with a unique opportunity to mark the passage of time. The epistemological basis of historical analysis is the same regardless of whether the subject of study is organisms or artifacts. Analytical methods are the same as well. Although the methods have precedence in the social sciences, they were first worked out explicitly in biology, and thus we examine them in that context before turning to archaeological applications.

We begin with what we have termed phyletic seriation (Fig. 3.1), as it involves the most commonsensical way to measure the similarity of objects and to order them based on their similarities. Heritable continuity, and thus historical continuity, is represented by a line of forms believed on theoretical grounds to comprise a lineage. Each form in the line is succeeded by a slightly different form, one that differs in one or several character states, such that change occurs continuously and gradually over time. Once this is understood, we can examine how heritable continuity should be evaluated analytically and explore the pitfalls that await the analyst.

CONTINUITY AND THE STUDY OF ORGANISMS

One axiom that biology students are taught early on is that natural selection works on individual organisms but that it is species, not individuals, that evolve. This might be confusing since species comprise individuals, and it is at the level of the individual that change shows up over generations. Further, we can see individuals, but we cannot see species. How do we know when a species has evolved? It would seem that this is an important question relative to keeping track of time, since we could mark the passage of time by identifying the appearance of new species. The answer is beyond the scope of this book, but suffice it to say that biologists themselves wrangle over the definition of a species and how to recognize one. Some fall back on the biological species definition; namely, that a species is a community of populations, the members of which interbreed or potentially interbreed under normal circumstances. As biologist Ernst Mayr (1980:34) put it, the "major intrinsic attribute characterizing a species is its set of isolating mechanisms that keeps it distinct from other species."

Paleontologists, however, have no way of determining, for example, whether the trilobites they are studying could have interbred with trilobites a few centimeters higher or lower in a stratum; in other words, they cannot tell how many biological species might be represented and they usually cannot tell much about the isolating mechanisms that were operative 450 million years ago. In short, they do not find the concept of the biological species very useful. This is how paleontologist George Gaylord Simpson (1943:171) defined the problem:

> [A] species as a subdivision of a temporal, or vertical, succession is a quite different thing from a species as a spatial, or horizontal, unit and it cannot be defined in the same way. The difference is so great and, to a thoughtful paleozoologist, so obvious that it is proper to doubt whether such subdivisions should be called species and whether vertical classification should not proceed on an entirely different plan from the basically and historically horizontal Linnaean system.

Note that Simpson was not denying that species were real; rather, he did not see the value in spending much time worrying about the problem. To him, Mayr's biological species, what Simpson termed a "spatial or horizontal unit," was one thing; Simpson's own paleontological species, a "subdivision of a temporal or vertical succession," was another. Simpson (1951:286) later labeled as "arbitrary" his procedure for breaking a temporal succession of similar fossils into species and characterized it as comprising situations "when organisms are placed in separate groups although the information about them indicates essential continuity." Criteria used in drawing arbitrary boundaries included stratigraphic as well as morphological discontinuities or changes in a few character states (Simpson, 1943, 1951). Resulting groups, such as the species of trilobites shown in Fig. 3.2, were not "artificial or unreal: they are natural groups approximating populations that once existed in nature" (Simpson, 1943:176).

Simpson (1943:174) captured the notion well when he indicated that the heritable continuity represented by a temporal sequence of genetically related forms—a lineage—comprised a "chronocline." Some paleobiologists now refer to the arbitrary chunks of a chronocline as "chronospecies," or "a segment of a lineage judged to encompass little enough evolution that the individuals within it can be assigned a single species name. In practice, a typical chronospecies does not exhibit a great deal more total variability, from end to end, than is found among the living populations of a single species" (Stanley, 1981:14; see also Stanley, 1979:13, 64). "Chronospecies, by definition, grade into each other, and each one encom-

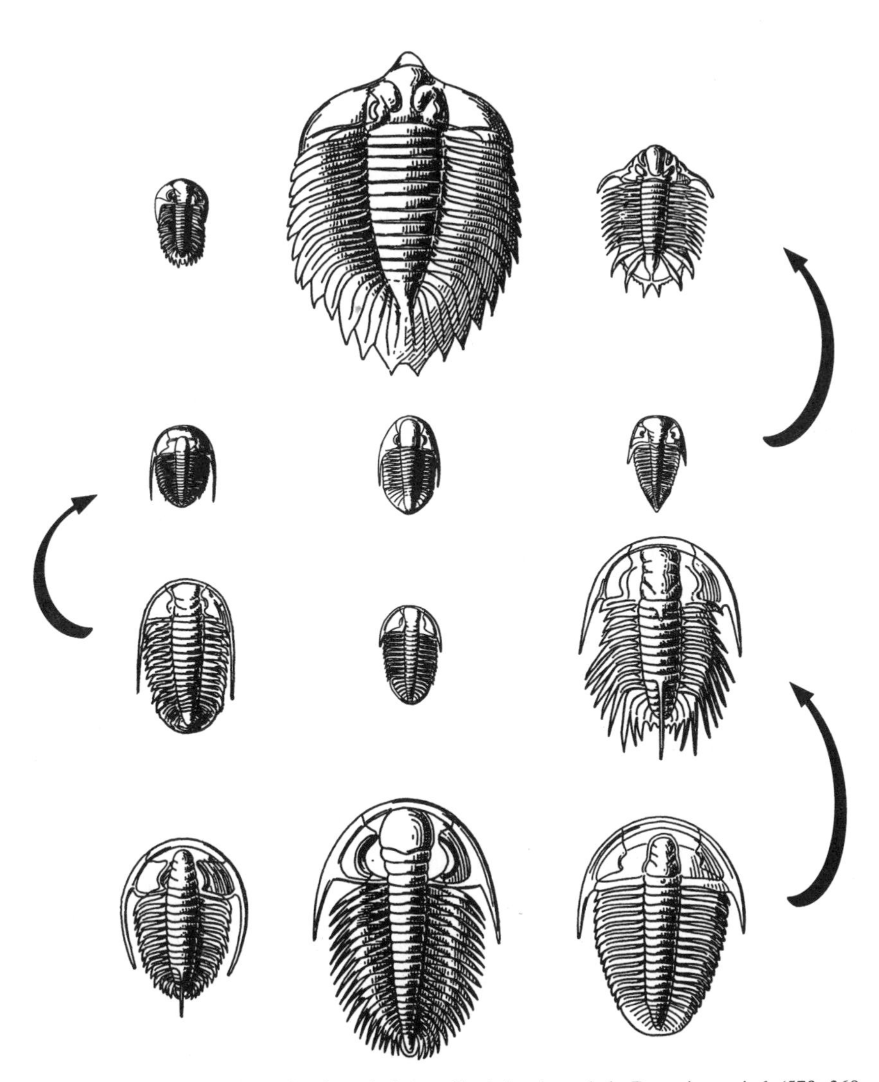

Figure 3.2. Trilobite species dating from the Lower Cambrian through the Devonian periods (570–360 million years ago). The oldest specimen is in the bottom-left corner, and the youngest is in the upper-left corner; arrows show direction of decreasing age. Trilobites are extinct members of the phylum *Arthropoda*, which includes organisms with segmented bodies and appendages. In general, the evolutionary trend was toward a reduction in the number of thoracic segments (those in the middle of the body), an enlarging of the pygidium (the posterior), and a change from spinous extremities to bluntly rounded ones. Middle specimen in the top row is 14.1 cm long (after Moore *et al.*, 1952).

passes very little change" (Stanley, 1981:93). Importantly, because character gradients comprise the lineage, "taxonomic division of a continuum into chronospecies is necessarily subjective and arbitrary" (Stanley, 1979:13).

Simpson would point out that regardless of how one breaks the continuum up, the resulting chunks will contain populations of organisms. Maybe they will not contain all the organisms that made up a population; in fact, they rarely will, but they will contain enough organisms so that a population is approximated. To Simpson, this population was a species. When one went high enough in a column of sediment that the organisms began to look different enough from the ones below in terms of morphology, truncate the first species and start a new one. In some cases, stratigraphic disruptions might provide some guidance in where to draw a boundary. Better yet, morphological change might correlate positively with stratigraphic boundaries, thus providing a double reason for stopping one species and beginning another. But, and this is an important but, such boundaries really are unnecessary. When fossils look different enough from those above or below, chop the sequence into segments. Those segments are theoretical units that contain other theoretical units, namely the species or chronospecies themselves. The empirical units are the specimens placed in each species based on shared properties. Does this not sound a lot like the advice offered by Phillips, Ford, and Griffin (1951) on how to create archaeological types (see Chapter 2)?

Figure 3.3, which comes from paleontologist Philip Gingerich's (1976) work in early Eocene beds in the Bighorn Basin of Wyoming, illustrates Simpson's position. Geological beds are shown on the vertical axis, along with a thickness scale. The horizontal axis measures the logarithm of the length times the width of the first lower molars of the sample of specimens that taxonomically fall in the primate family Adapidae. The horizontal dashed lines represent Gingerich's species boundaries in terms of tooth size—the character of interest—and where in the column the (chrono)species happen to fall. Note that, allowing for minor variation, there is a continuous, gradual change toward larger tooth size from the bottom of the sequence to level 1100, when tooth size diversifies (level 200 to level 540 represents about 1–1.5 million years) (Gingerich, 1979; Gingerich and Simons, 1977). Notice what Gingerich said about change through time and how he defined his species:

> Change in *Pelycodus* [the genus shown most often in Fig. 3.3] is both continuous and gradual. Evidence that the change is continuous and gradual is given by the fact that wherever the record is sufficiently dense, there is no statistically significant difference between adjacent samples. Over large intervals, however, significant differences do accrue as the sum of numerous insignificant differences between adjacent levels. Hence the sample of specimens labeled *Pelycodus trigonodus* in [Fig. 3.3] is recognizably different, even to the unaided eye in the field, from most samples of *P. mckennai* or *P. abditus*, even though intermediate samples connect them all into one continuous temporal gradient. The result is a continuous evolutionary lineage, subdivided arbitrarily into the segments *Pelycodus mckennai*, *P. trigonodus*, *P. abditus*, etc. (Gingerich, 1979:48)

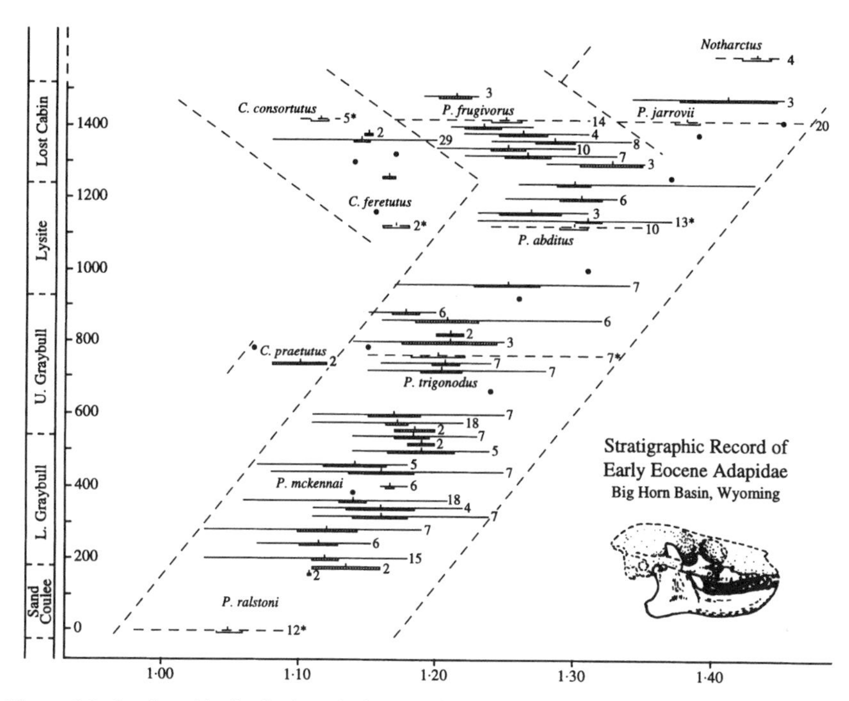

Figure 3.3. Stratigraphic distribution of primate *Pelycodus* and the related genus *Copelemur* in early Eocene sediments of the Big Horn Basin, Wyoming. Horizontal axis is tooth size (the log of length times width of the lower first molar) and, by inference, body size. Names along the vertical axis are formations and numbers are stratigraphic position (in feet) above the base of the Willwood Formation. Horizontal line is sample range; vertical slash is mean; solid bar is standard error of mean; small number is sample size. Dashed lines show the pattern of linking for species of North American *Pelycodus* (after Gingerich and Simons, 1977).

A slightly different way of looking at the relation between a (chrono)species defined on paleontological evidence and the biological notion of a species is shown in Fig. 3.4. Species as groups of related, potentially interbreeding individuals, change morphologically through time. Species are also parts of lineages. If enough time has passed since lineages diverged, species in two different lineages will be readily distinguishable. Thus at time t_2 in Fig. 3.4, there are nonarbitrary discontinuities between species A and B; likewise, at time t_3 there are similar discontinuities between species A and C. In contrast, the distinction between two successive (chrono)species—B and C in Fig. 3.4—is arbitrary in cases where the fossil record is sufficiently complete to show the transition from one species to another. As Gingerich (1979:48) well put it,

> This need for an arbitrary boundary does not mean that differences between two successive species do not exist. Rather, the problem is analogous to one of keeping time.

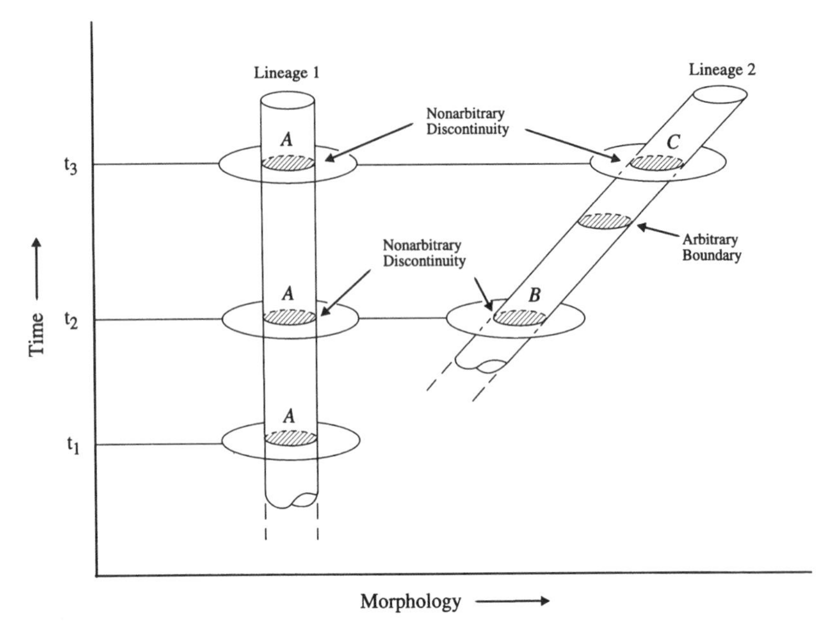

Figure 3.4. Diagram showing the relation between biological species (e.g., A and B) sampled at any given time *t* and generalized paleontological species viewed as units of evolutionary lineages (e.g., B and C). Species in the fossil record have both a biological dimension, where nonarbitrary morphological discontinuities are used to infer reproductive isolation on any given time plane, and a time dimension, where successive morphologically changing units of an evolving lineage must be separated at an arbitrary, time-parallel boundary. In some instances, lineage branch points furnish natural, nonarbitrary boundaries between successive units of evolving lineages (after Gingerich, 1976).

> The fact that one o'clock grades continuously into two o'clock does not mean that the two hours are the same; we make a necessarily arbitrary boundary between them and recognize that they are different.

This is time's arrow, or time as a continuum, sliced arbitrarily by the notations one o'clock, two o'clock, and so on into units.

Tracing Lineages

Day-to-day changes in the composition of the species rendered as turnover in the individuals comprising its population(s) not only drives the evolution of that species but leaves traces, manifest as morphological variation, that allow us to reconstruct its evolutionary pathways or lineages. Piecing together lineages, measuring rates of evolutionary change, and trying to understand the processes that create change are part and parcel of the world of paleobiology. The specific methods and techniques that have been devised to trace historical lineages are anything but simple, but at their core revolve around identifying similarities and differences in character states of organisms. A character state is the condition or variable state (sometimes termed an attribute) of a character or dimension of variation. At one particular time all organisms in a species might have, for example, five digits, but in a succeeding generation some might have six. Such constitutes a change in the state of the character "number of digits." This simple example should make it clear that measurable changes in characters, a shift from one state to another, are useful for documenting the passage of time. Given enough changes and if those changes result in reproductive isolation, meaning that the organisms with the changes begin to interbreed only among themselves, it is possible that a new species will be formed alongside the ancestral species.

But we are not particularly concerned with change at the species level. Rather, we are interested in creating an order of specimens that, when viewed overall, as in Fig. 3.3, measures the passage of time. This is about as close as we can come to measuring time continuously rather than as a series of chunks. If we pluck out a specimen from the continuum and turn it into an archetype, meaning that we use it as a representative specimen or average (norm) of numerous specimens, then we have created an index fossil. Its main uses are to correlate in time spatially distinct phenomena and to measure the passage of time in discontinuous chunks (see Chapter 6). The important point here is that we do not have to use stratigraphic or other evidence to develop a temporal sequence of variant forms. We can use historical continuity, measured in terms of how similar in form variants are to one another in terms of their shared character states, the product of heritable continuity. The second point is where potential problems arise.

Intuitively, the more similar two things are to each other, the more character states they hold in common, the more closely related they probably are. Conversely, the more dissimilar they are, the more distantly related they probably are. A more appropriate expression is to say that because two things are closely related they thus are similar. As Simpson (1961:68–69), put it, monozygotic twins are not twins because they're identical. Rather, they are identical because they are monozygotic twins. With regard to organisms, similarity is a result of genetic replication; with artifacts, it is a result of nongenetic replication (Leonard and Jones, 1987). The important point is that it is replication. Regardless of whether we are dealing with organisms or clay cooking pots, the process of replication is plainly evident, and in both cases it involves transmission. In organismic reproduction, organisms transmit genetic material through asexual or sexual reproduction, creating either an offspring that is an exact copy of the parent (asexual) or an offspring that has characteristics of both (sexual). In both cases, reproduction is more or less faithful, meaning the offspring is very similar or identical genetically and phenotypically to the parent or parents. In nonorganismic reproduction, an organism or a group of organisms manufactures objects that usually are more or less faithful replicates. Transmission is as important here as it is in organismic reproduction. For example, a parent teaches a child how to manufacture and decorate a cooking pot, the end result being the production of pots that look very similar to those produced by the parent. Over time, because of either transmission errors or innovation, pots change in terms of form and/or decorative motif. These changes might not be detectable on even a yearly basis, but after a time, if we looked at the pots in production now and compare them to pots made, say, 30 years ago, we would probably notice significant variation. In other words, the two ends of the chronocline would comprise obviously dissimilar specimens, whereas specimens adjacent to one another in the chronocline would be virtually identical.

The same is true for species and the individuals that compose them. Unless we are dealing with organisms with incredibly rapid reproductive rates, and here only in a laboratory situation, we can seldom see species evolve. But we can take careful measurements of individuals and keep track of the changes that occur in character states across successive generations. It is precisely these changes that help us keep track of time. Paleontologists and archaeologists, however, do not deal with sequences of genetically related, ancestral and descendant living organisms. Paleontologists deal with the hard parts of once-living organisms—fossils—and archaeologists deal with the products and by-products—artifacts—left by humans. Just as paleontologists can track individual variants through time, meaning they track the production of new variation as it becomes manifest in individuals, so too can archaeologists track the emergence of new variants. And they can rely on the similarity of specimens to create a chronological order by accepting the assumptions of historical and heritable continuity.

Detecting Heritable Continuity

Earlier we noted that it seemed reasonable to suppose that the more similar two things are to each other, the more closely related they probably are, and the greater their dissimilarity, the less closely related they are. This is simply another way to state the basic principle of heritability, or Darwin's "descent with modification," and so it approximates the assumption of heritable continuity underpinning the seriation method. But concluding on the basis of similarities that two phenomena are related does not necessarily mean that we are dealing with temporal as well as formal or morphological differences. Let us assume, for example, that we have samples of fossils of two contemporaneous congeneric species, such as species A and B at time 2 (t_2) in Fig. 3.4. In the absence of independent chronological data such as might be provided by superposition, these species could be seriated into an order such that one preceded the other, but this ordering would not provide an accurate measurement of the relative age of the fossils because the two are contemporaneous at t_2. That lineage 2 in Fig. 3.4 diverged from lineage 1 and one of the former's products was species B while the product of the latter continued to be species A conflates the assumptions of historical continuity and heritable continuity. To produce a correct chronological ordering of the fossils we have chosen to variously place within the categories species A and species B, we must disentangle empirical manifestations of the two assumptions.

To assess relatedness, biologists use the states of anatomical characters—character states—such as tooth number, tooth size or shape, or pelt color. But they also do something else that is important: They recognize two broad classes of characters—analogous characters and homologous characters. The former are characters that two or more organisms possess that, while they might serve similar purposes, did not evolve because of any common ancestry. One common classroom example is the wing. Birds and bats both have wings, and those characters share properties in common, yet we classify birds and bats in two widely separate groups. Why? Because birds and bats are only distantly related; those two large groups diverged from a common vertebrate ancestor long before either one of them developed wings. Thus we say that bird wings and bat wings are analogous structures, or analogues for short. They are of no utility in reconstructing lineages because they evolved independently in the two lineages after they diverged. The character of having wings is held in common by birds and bats, but the state of the character, the details of its osteological composition and anatomical structure, differs between the two groups.

Analogous characters and character states are of no use in tracking heritable continuity but homologous characters, or homologues for short, are. Homologues are used for this purpose because they are characters and character states that are primitive holdovers from the time when two lineages were historically a single lineage. Think of the similarities that we share with the great apes (gorillas, chimpanzees, and orangutans), the living primates with which we are most closely related historically and genetically, or in terms of heritable continuity. Based on the sizable number of characters or character states that we share with chimps, we might reach the conclusion that they are our closest living relative, and we would be correct, as shown by DNA comparisons. In essence we are saying that because we are so similar to chimps, we and they must be fairly closely related phylogenetically. Again, what we ought to be saying is that humans and chimps are similar because they are related.

Merely identifying the distinction between analogous and homologous characters, however, in no way ensures that they can always be distinguished among fossils or organisms. We reasonably expect that the preponderance of characters that humans and chimps possess in common are homologous, but there is no reason to suspect that some shared characters could not have been derived independently by both chimps and hominids (the lineage that includes at least the genera *Australopithecus* and *Homo*) after their divergence some 6 million years ago. Dozens of methods exist for making the distinction between analogous and homologous characters, including analysis of detailed morphological characteristics to determine whether they are structurally similar or only superficially similar.

Ironically, the homologous characters themselves cause the greatest problem.

They are, in effect, double-edged swords, on the one hand presenting us with the means of tracing heritable continuity, and thus relatedness, and on the other laying traps that are easy to walk into. For example, all mammals have a vertebral column, as do animals placed in other categories, such as most fishes. The presence of vertebrae is one of the criteria that we use to place organisms in the subphylum Vertebrata. The vertebral column is a homologous character shared by mammals and fishes, but it is a character that goes so far back in time as to be essentially meaningless in terms of helping us understand how the myriad backboned organisms of the last 400 million years are related phylogenetically. Thus we use other characters such as the presence or absence of hair or a four-chambered heart to segregate mammals from other classes of organisms that have backbones. This segregation—cut, for short—took place about 200 million years ago. Then we make another cut based on the presence–absence of other characters to subdivide the sample further, then another cut, and another, and so on. We use what are called shared derived characters, or synapomorphies, to do this; shared primitive characters, or symplesiomorphies, are not considered. The latter characters, such as the vertebral column, are indeed homologous, but they do not help in the construction of phylogenies precisely because they are shared by all members of all the groups of Vertebrata.

The basic analytical procedure is not as straightforward as one might hope. Figure 3.5 illustrates three possible historical patterns (two are termed *phenograms* and one is termed a *cladogram*) for four taxa (A–D) and five characters (1–5). As shown in the matrix at the lower left, for each taxon the five characters are in one of two states, designated by a lowercase letter with and without a prime. The matrix in the lower middle shows the number of shared character states between pairs of taxa. Which historical pattern is correct? Perhaps phenogram 1 is correct, since it denotes a close phenetic relationship between taxa A and C, which have four character states in common, but then again, it splits out taxon D, which has three states in common with taxon A.

Perhaps phenogram 2 is better in that it minimizes the phenetic distance between taxa D and C and shows that taxon A is closer to D and C than any of them is to taxon B. On the face of it, phenogram 2 is far superior to phenogram 1, a result that we probably would have gotten by plugging the data into most cluster analysis programs. Such algorithms search all the data to find the most equitable solution in terms of minimum–maximum distances (similarity rendered as shared attributes) between pairs. But does phenogram 2 configure the historical evolutionary relationships among the four taxa correctly? We would always bet heavily against the correctness of any phenogram that purportedly illustrates phylogenetic relationships. We would make this bet because in the Fig. 3.5 example, as is true of most phenograms, we made no attempt to discriminate among the kinds of homologous features used in the analysis. Instead, we lumped synapomorphies and symplesiomorphies together.

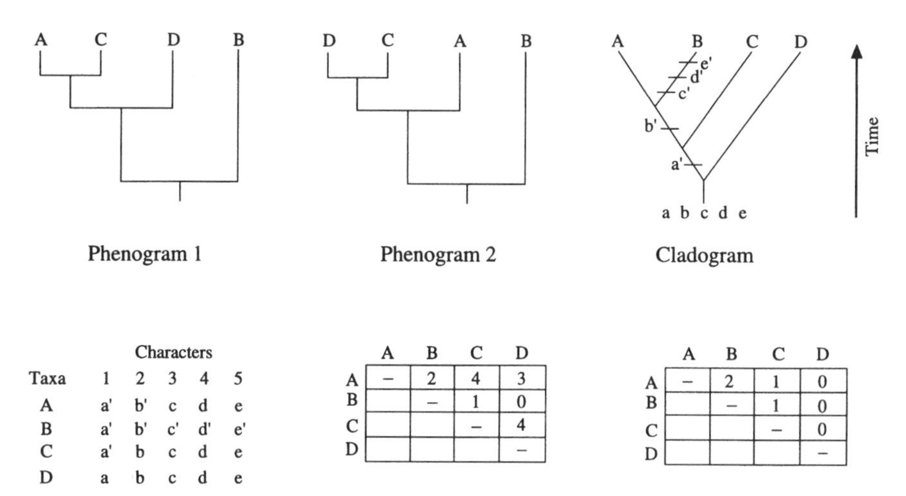

Taxa	Characters 1	2	3	4	5
A	a'	b'	c	d	e
B	a'	b'	c'	d'	e'
C	a'	b	c	d	e
D	a	b	c	d	e

	A	B	C	D
A	–	2	4	3
B		–	1	0
C			–	4
D				–

	A	B	C	D
A	–	2	1	0
B		–	1	0
C			–	0
D				–

Figure 3.5. Comparison of two phenograms (upper left and middle) with a cladogram (upper right), based on the hypothetical states of five characters in four taxa (lower-left matrix). The phenograms are constructed from the total number of character states shared by any pair of taxa (lower-middle matrix), whereas the cladogram is constructed from only the derived character states (marked by primes) shared by pairs of taxa (lower-right matrix). The character state transitions are marked on the cladogram. Note that the evolutionary tree implied by the use of shared derived characters differs from those implied by the use of all shared characters (after Futuyma, 1986).

The pattern at the upper right in Fig. 3.5—the cladogram—illustrates the evolutionary history of the four taxa based solely on an analysis of shared derived characteristics, shown in the matrix at the lower right. In other words, here we know what the primitive state of each of the five characters is—lower case letters without a prime—and we ignore them in favor of the derived states—lowercase letters with a prime. Now taxa A and C share little in common, whereas before they were placed closer together because of the total number of character states they had in common. But three of them were primitive states, not derived states. Analysis of synapomorphies alone indicates that taxon C split off from the ancestral form that produced A and C at some point when taxon B had not yet come into existence. Cladistical analysis produces what are called *monophyletic groups*, or groups that include all descendants of a particular ancestor plus that ancestor. In the Fig. 3.5 cladogram there are three monophyletic groups, or clades (not counting single taxa): AB, ABC, and ABCD (each would also include the here-unidentified common ancestor). Any other group, ACD, for example, is polyphyletic, and thus is disallowed.

Strict adherence to using only synapomorphies to determine phylogenetic history is suggested by some (e.g., Cracraft, 1981) but certainly not by all biologists (e.g., Bock, 1977; Mayr, 1981). The latter maintain that cladistical analysis is not the only means of assessing phylogenetic history or relatedness and is not the best method because it ignores the temporal dimension. Some who use cladistic analysis maintain that there needs to be an independent evaluation of lineages created from, say, stratigraphic interpretation and that cladistics offers a means of doing just that (Eldredge and Novacek, 1985). Others argue the opposite and suggest that cladistically based interpretations should be tested with stratigraphically superposed fossils (Smith, 1994). The emergence of stratocladistics (Fisher, 1994) is one expected result, as each method serves to test and strengthen inferences founded on the other. But whether the temporal dimension is used during the process of building phylogenetic histories or to test such histories, cladistical analysis, given its focus on synapomorphic characters, is a critically important procedure.

There is a subtle twist to cladistics, and it has to do with time. Some cladists argue that time is irrelevant to what they do and that superposed fossils may be misleading if the fossil record is differentially complete, mixed, or temporally inverted. What they mean when devaluing the temporal dimension is that a cladogram tells us nothing about how much time has elapsed between branching episodes. That's why cladograms do not reflect differences in amounts of time but rather only a relative temporal sequence of branching events. Just by looking at the cladogram in Fig. 3.5, for example, we do not know what the length of time was between the point that taxon D left the ancestral group and the point that taxon C left it. Nor do we know whether that length of time was greater than, equal to, or less than the amount of time that passed before taxa A and B diverged. All we know is the sequence of events. But this sequence itself marks the passage of time in a relative sense.

Figure 3.6 illustrates another cladogram, this one of dinosaurs. The presence of a hole in the hip socket is one characteristic used to classify organisms as dinosaurs. But being a primitive characteristic shared by all dinosaurs—a symplesiomorphy—it is of no use in illustrating ancestral–descendant relations among the dinosaur taxa. One derived characteristic is the presence of a backward-pointing extension of the pubis (Fig. 3.6); it is used to separate theropods and saurischians into one group and ornithischians into another. The ornithischians can be further divided on the basis of layers of tooth enamel. Stegosaurus, for example, has equal layers of enamel, whereas the other three taxa shown in Fig. 3.6 exhibit unequal layers, another derived characteristic. Marginocephalians can be further distinguished from the ancestral group by the presence of a bony shelf at the back of the skull, another derived characteristic.

Were character states replaced at a constant rate during biological evolution, it would be a bonanza for those of us worried about measuring the passage of time on a calendric scale. Unfortunately this is not the case. As an example, consider chimpanzees and humans. We resemble chimps in several morphological respects,

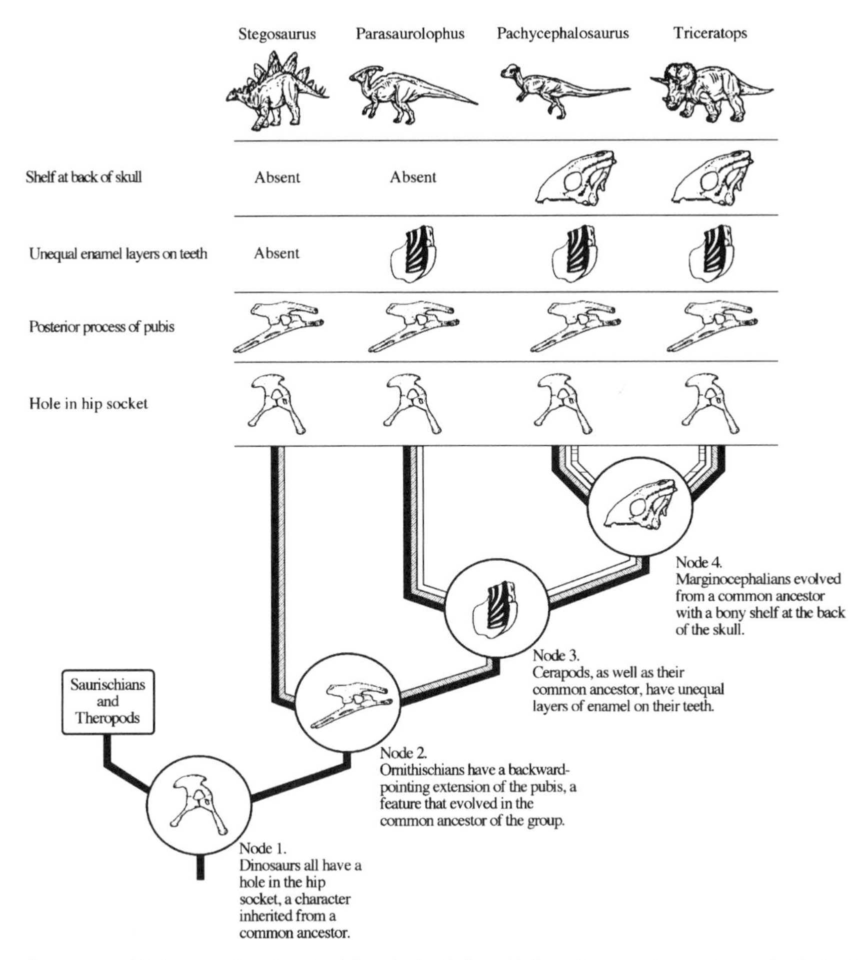

Figure 3.6. Cladogram showing the historical relation of four dinosaur taxa using only derived characteristics. All dinosaurs have a hole in the hip socket; this is one criterion we use to classify organisms as such. At node one, the saurischians and theropods have begun to diverge from the ornithischians, the latter of which have a backward-pointing extension of the pubis, a derived characteristic that evolved in the common ancestor of that group. Saurischians and theropods do not have the extension. The ornithischians can be further divided on the basis of layers of tooth enamel: Stegosaurus has equal layers, where the other three taxa shown, the cerapods, have unequal layers, another derived characteristic. Marginocephalians can be distinguished from the ancestral cerapod group on the presence of a bony shelf at the back of the skull, another derived characteristic (after Gaffney *et al.*, 1995).

but we also differ from them in many other respects, so many that chimps are placed in a genus, *Pan*, separate from our own, *Homo*. In terms of DNA, however, humans and chimps are extremely similar, suggesting that morphological and biochemical evolution have proceeded at different rates (King and Wilson, 1975). There really is no reason to think that evolutionary rates should be similar for morphological characters and biochemical characters. True, all morphological characters are ultimately the results of biochemical characters, and if we want to be specific, we should say that some biochemical characters are much more conservative than others. In the grand scheme of things, conservatism and stability seem to be typical for species. To be sure, changes in the genetic and morphological makeup of the organisms within a species take place constantly, but these changes often are almost undetectable. It is the cumulative effect of the changes, which took place over long spans of time, that we see in the fossil record. We return to the issue of deriving absolute dates from seriations in Chapter 4.

The assumptions of historical continuity and heritable continuity, though related in important ways, are separate and distinct. Historical continuity denotes a sequence of similar forms that measures the passage of time; heritable continuity underpins the assumption of historical continuity when the latter is rendered as formal similarity and used to order phenomena in what is inferred to be a temporal sequence. Heritable continuity, or "relatedness," is one explanation for the observed, historically significant similarities. But similarities can arise as a result of heritable continuity, yet arise in such a manner as to not measure the passage of time (e.g., species A and B at t_2 in Fig. 3.4), or they can arise as a result of evolutionary convergence that produces analogous character states. As we will show in the following, archaeologists early on were well aware of precisely these sorts of problems, but they chose to deal with them analytically in quite different ways than paleobiologists. Some archaeological approaches to distinguishing between analogous and homologous similarity were reasonable, some were not. More often than not, if the distinction was noted by archaeologists at all, it was superficial notice. Most importantly, the general failure of archaeologists to distinguish conceptually between historical and heritable continuity resulted in them variously conflating the two at one extreme and allegedly discarding the latter while retaining the former at the other.

HISTORICAL CONTINUITY, HERITABLE CONTINUITY, AND THE STUDY OF ARTIFACTS

Archaeologists have long used changes in artifact form rendered as changes in character states to measure the passage of time, the idea being that stone tools or ceramic vessels, for example, evolve in terms of form. Figure 3.7 illustrates James A. Ford's (1962) conception of culture change as seen in vessel form. Such a neat, orderly, and gradual evolution of form through time could only be reconstructed through detailed and painstaking research. Almost from the beginning, American-

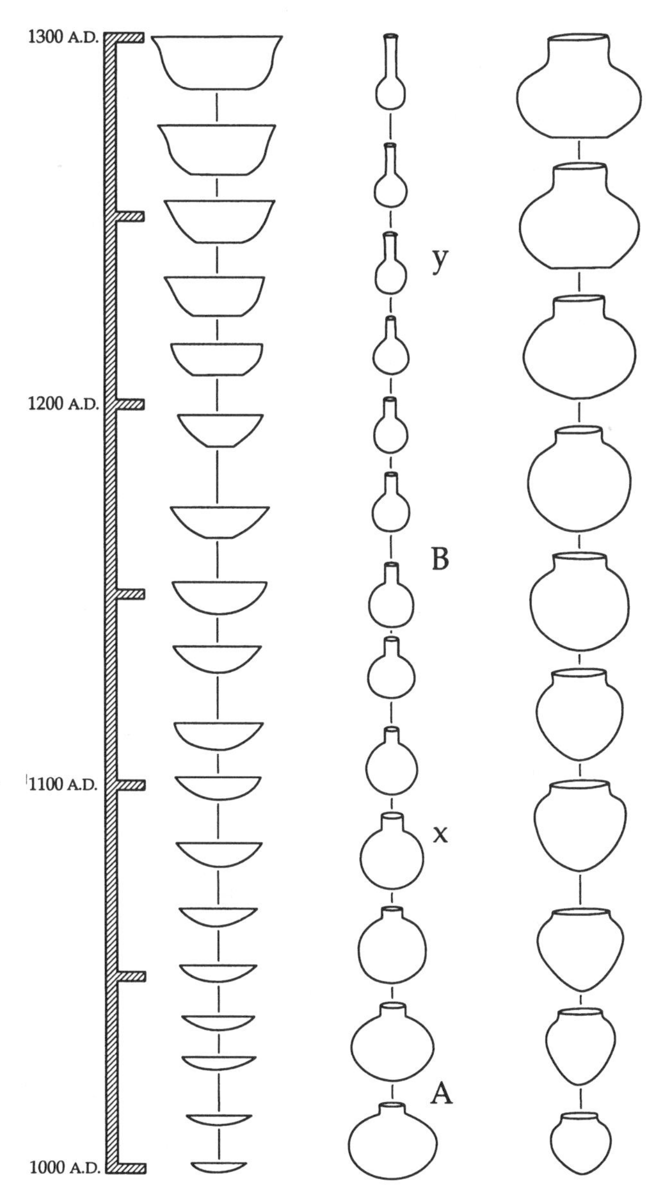

Figure 3.7. James A. Ford's conception of culture change as exemplified in pottery. Note the constant gradation in the three forms through time. Ford also used the diagram to reinforce the arbitrariness of types. If a type A is set up in the water bottle tradition as shown, then the classifier has to select the next type in this tradition at least as far away as B in order to differentiate among types consistently. The examples midway between these "typical" type examples are the difficult borderline cases. Alternatively, if the first type established had been set up at *x*, then the second type would have to be as far away as *y*. In this arrangement, the forms that were typical for type B of the first system become the doubtful specimens of the second (after Ford, 1962).

ist archaeologists looked to their sister disciplines of geology and paleontology for guidance in how to measure time. The notions they borrowed from geology—superposition and stratigraphy—were straightforward enough, but those borrowed from paleontology and biology were anything but straightforward. Archaeologists treated biological concepts in a casual fashion and in some cases made it sound as if artifacts were capable of breeding and producing offspring. When archaeologists interested in applying concepts of biological evolution to the study of culture change were charged in the 1940s with being naive reductionists, they abandoned any effort to incorporate biological theory into the study of archaeological change, and hence the study of time.

Culture Traits

Early in the twentieth century ethnologists attempted to measure time with culture traits and the age–area concept (see Kroeber, 1931b, and Sapir, 1916, for reviews). This concept was built on three interrelated assumptions: (1) Culture traits will disperse in all directions from the point of origin, like the ripples emanating from a rain drop that hits a puddle; (2) all culture traits will disperse at the same rate, again, like ripples; and (3) the larger the geographic area over which a culture trait is found, the older that trait is. Thus, mapping the geographic distribution of culture traits evidenced by archaeological remains should indicate something of their history and the passage of time. The age–area concept was supported in some quarters of anthropology and archaeology (e.g., Kroeber, 1931a,b) because an identical concept existed in biology (e.g., Willis, 1922), though it was by no means completely accepted there. Not surprisingly, then, criticisms of the age–area concept appeared (e.g., Steward, 1929). The concept did not recognize independent invention or convergence, processes that everyone agreed took place even if their relative importance might be disputed. The age–area concept presumed that similarities in different cultures, that is, the lists of traits comprising cultures, were homologous (the result of shared ancestry) and largely ignored the potential of analogous similarities. One anthropologist in particular outlined the solution to precisely these sorts of problems. The source of his solution ultimately resided in the theory of biological evolution.

A. L. Kroeber (1931a) argued that culture traits were analogous to species of organisms and that cultures, which comprise suites of traits, were analogous to faunas and floras, which comprise suites of animal and plant species, respectively. For Kroeber (1931a:149), although a "culture complex is 'polyphyletic' [and] a genus is, almost by definition, monophyletic ... the analogy does at least refer to the fact that culture [traits] like species represent the smallest units of material which the historical anthropologist and biologist respectively have to deal with." Therefore, data on the geographic distribution of culture traits, as with such data on plant and animal taxa, allow "inferences as to the origin and areal history of the group" because the "Age and Area principle seems the same in biology and cultural anthropology" (Kroeber, 1931a:150). Kroeber was wrong when he said that species are the smallest units that biologists deal with because they deal also with genes and also the organelles in cells, both of which were well-known biological phenomena when Kroeber made his comment. Biologists do, of course, study the individual organisms and populations thereof that comprise a species. We suspect Kroeber chose to equate species with culture traits because he was thinking that both sorts of units evolve monophyletically, and thus could not be reduced to smaller evolutionarily significant units, whereas cultures, as he noted, are polyphyletic (being comprised of a constellation of traits). But Kroeber left the critical term "culture trait" undefined, as did most of his anthropological and archaeological colleagues. Most importantly, at what scale should a culture trait be conceived? As a combination of particular discrete objects arranged in a particular structure? As a discrete object? As an attribute of a discrete object? Failure to consider the issue of scale was ultimately a serious problem, as we will see.

Kroeber, however, well understood the difference between analogues and homologues:

> The fundamentally different evidential value of homologous and analogous similarities for determination of historical relationship, that is, genuine systematic or genetic relationship, has long been an axiom in biological science. The distinction has been much less clearly made in anthropology, and rarely explicitly, but holds with equal force. (Kroeber, 1931a:151)

He implied that a "true homology" denoted "genetic unity," arguing that

> There are cases in which it is not a simple matter to decide whether the totality of traits points to a true [genetic, homologous] relationship or to secondary [analogous, functional] convergence.... Yet few biologists would doubt that sufficiently intensive analysis of structure will ultimately solve such problems of descent.... There seems no reason why on the whole the same cautious optimism should not prevail in the field of culture; why homologies should not be positively distinguishable from analogies when analysis of the whole of the phenomena in question has become truly intensive. That such analysis has often been lacking but judgments have nevertheless been rendered, does not invalidate the positive reliability of the method. (Kroeber, 1931a:152–153)

Kroeber was suggesting that there are two forms of similarity: one homologous and the other analogous. The former results from shared genetic ancestry; the latter results from evolutionary convergence, such as when two genetically unrelated populations of organisms reach a similar adaptive solution. How are the two distinguished? Kroeber (1931a:151) suggested that "Where similarities are specific and structural and not merely superficial ... has long been the accepted method in evolutionary and systematic biology." Kroeber also recognized the problem of independent invention—analogous similarity—and noted that anthropologists had too often not ascertained if the traits they deemed similar were the result of common ancestry—homologues—or the result of evolutionary convergence—

analogues. Sapir (1916) had spelled out in general terms the analytical procedure and criteria for rendering such ascertainment, and they were summarized in detail by Julian Steward in 1929.

When a cultural trait was found in two or more localities, the criteria were: (1) the "uniqueness" or quality of the cultural trait; (2) the presence of a probable ancestral trait in the same geographic area; (3) the quantity of other shared traits; and (4) the geographic proximity of the localities (Steward, 1929). Only by close study of each of these criteria could one determine whether a cultural trait in an area had originated there or elsewhere. The uniqueness criterion was often expressed as a trait's complexity; the more complex a trait, the less likely it was independently invented multiple times and the more likely it was a homologue. The issue of scale—was the shape of a decorative design important, the elements and color(s) it comprised, its placement on a vessel, or some combination of these representative of a trait's complexity—never arose. The quantity of shared traits and the geographic proximity of compared localities were thought to correspond directly to the probability of contact. The more traits shared and the closer in space the cultures sharing them, the more likely the traits were homologues. These three criteria were, then, the basis for the inference that similarity is of the homologous sort. Not surprisingly then, rather than indicate how ancestral traits were to be identified, Steward fell back on the other three criteria to help determine if an ancestral trait was present. This probably contributed to Kroeber's (1931a) lament two years later that anthropologists had failed to borrow biological procedures for distinguishing between homologous and analogous traits.

Early phyletic seriations of artifacts basically used Steward's criteria of spatial and temporal propinquity without explicitly acknowledging them. Similarity in artifacts was measured by examining the attributes of a category of artifacts such as pottery or coins or axes and noting how those attributes changed states. A coin did not become an axe, but a large coin might decrease in size over time. As the examples we now turn to demonstrate, the procedure of phyletic seriation was virtually identical to the technique of paleontologists, who arranged fossils of similar form in an order such that change in character states of the fossils was gradual and continuous.

W. M. Flinders Petrie and Artifacts from Egyptian Tombs

The earliest phyletic seriation using pottery of which we are aware was that carried out by British archaeologist Sir William Matthews Flinders Petrie, who was faced with making chronological sense out of some 4000 predynastic burials from several localities along the Nile River north of the Valley of the Kings in Egypt. Few were in superposed position (see Chapter 5), so that principle could not be used as a chronological indicator. Petrie (1901:4) mulled the problem over:

> [I]f we can use any definite scale of sequence, where the scale of absolute time is unknown, we can at once deal with a period as simply and clearly as if the scale of years was provided. Such a scale of sequence we have in the numbers of the burials; and if we can only succeed in writing down the graves in their original order of time, we can then be as definite in fixing their contents in a scale of graves as we would in a scale of years.
>
> The problem then is, if we have the contents of hundreds of graves accurately recorded, how can we sort those out into their original order, and so construct a scale?

Petrie (1899a:297) believed that artifacts could be ordered because they occurred in a "series of development and degradation of form"; after the initial ordering was produced, then the analyst could perform a "statistical grouping by proportionate resemblance" of various assemblages. His developmental sequences were constructed using *phyletic seriation*, which comprises the seriation technique for which he should be remembered. Once established, a phyletic sequence enabled "a long period to be arranged in approximate order, and serves as a scale for noting the rise or disappearance of other types" (Petrie, 1899a:297). The last, "noting the rise or disappearance of other types," was a tagalong benefit allowing "statistical grouping" of other assemblages "by proportionate resemblance"; thus, any resemblance of Petrie's procedure to the frequency seriation technique of Kroeber is fortuitous. Petrie clearly preferred phyletic seriation, though he did not use that term. He remarked, for example, "fibulae might be roughly classed by the proportions of pottery types that were found with them; but the similarities of form [of fibulae] would enable them to be put more exactly into order" (Petrie, 1899a:298). Further, although other assemblages might be added to the sequence based on the proportions of various artifact types they contained, such an addition took place *after* the sequence had been constructed on the basis of phyletic seriation. This is a decidedly different procedure than frequency seriation (see Chapter 4).

The drawings that accompanied Petrie's paper make it clear how the phyletic seriation technique worked. In Fig. 3.8, the sequence runs from oldest at the top to youngest at the bottom, though this is irrelevant. Petrie conceivably could have had the sequence reversed without disrupting the validity of the technique. In the diagram, Petrie illustrated various ceramic vessel forms that were in the sample of roughly 900 burials he used initially (he later added over 1000 more, once the sequence had been established). Vessel forms are arranged by period, from a point in the distant past, period 30, up to more modern times, period 80. Given that this is a much simplified drawing, Petrie combined many of the periods into larger units. Note that through time, some vessel forms continued through several periods, while others did not. For example, the outflaring cylindrical vessel form shown in the upper left continued from period 30 into period 34. The bowl form in the upper right did the same thing. This overlapping of forms from one period to the next allowed Petrie to work his way up or down through time to create a sequence. He later used a simple analogy to explain his method:

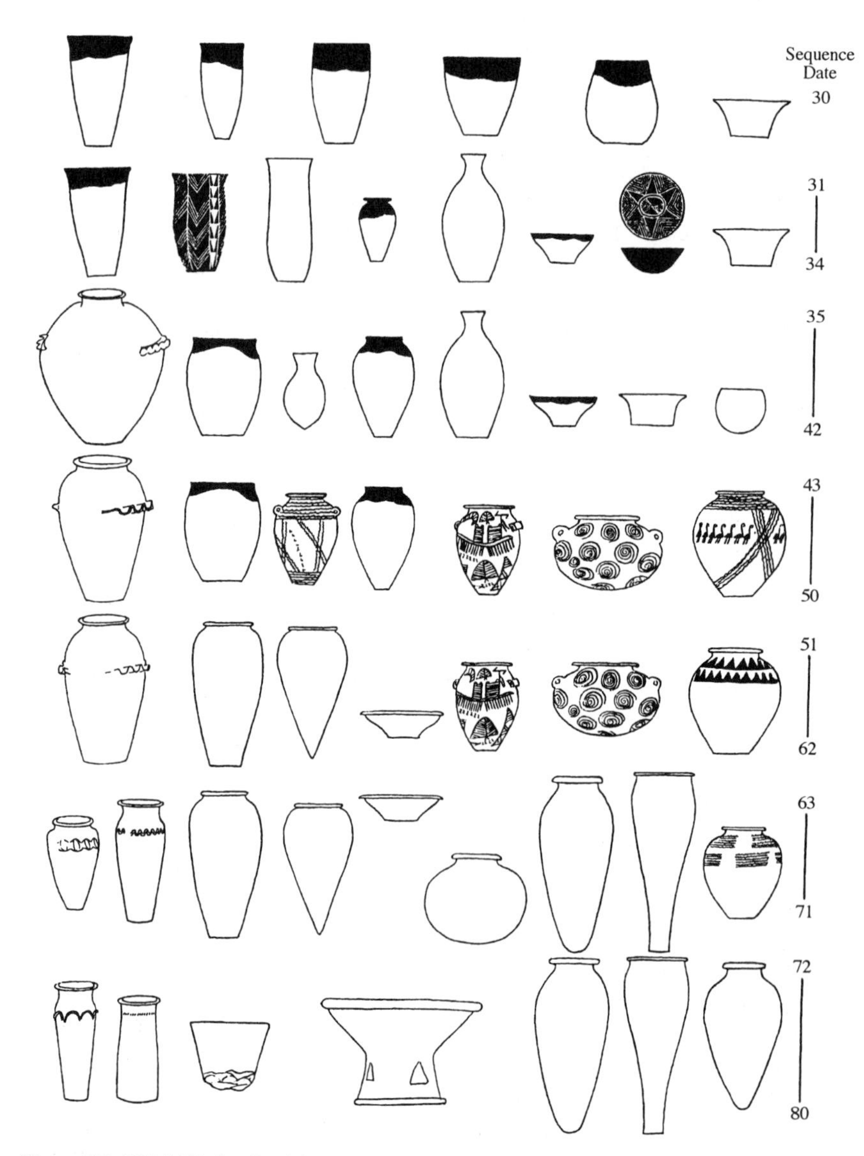

Figure 3.8. W. M. Flinders Petrie's chronological ordering of ceramic vessels recovered from burials in three localities in Egypt. Periods from 30 to 80 are shown at the right; vessels from each period are placed in rows. Notice the wavy handles on vessels at the far left of each row; the shape of the handle was the first clue Petrie had as to the sequential ordering of the burials (after Petrie, 1901).

> If in some old country mansion one room after another had been locked up untouched at the death of each successive owner, then on comparing all the contents it would easily be seen which rooms were of consecutive dates; and no one could suppose a Regency room to belong between [Queens] Mary and Anne, or an Elizabethan room to come between others of [King] George III. The order of the rooms could be settled to a certainty on comparing all the furniture and objects. Each would have some links of style in common with those next to it, and much less connection with others which were farther from its period. And we should soon frame the rule that the order of the rooms was that in which each variety or article should have as short a range of date as it could. Any error in arranging the rooms would certainly extend the periods of things over a longer number of generations.
>
> This principle applies to graves as well as to rooms, to pottery as well as to furniture. Having all our material so exactly denoted, according to so many minute varieties, we are able to frame extensive statistical tests for classifying it, and to deal with it by the arithmetical as well as the artistic arrangement. (Petrie, 1899b:26–27)

The key to Petrie's analysis was that he found an attribute that changed states through time, and thus allowed him to construct a temporal sequence of pottery forms. This attribute was vessel handles: "At the left ends of the five lower rows is the wavy-handled type, in its various stages; the degradation of this type was the best clue to the order of the whole period" (Petrie, 1899a:300). Petrie suspected that the handles were functional on the earlier jars, which tended to be large and bulky, but that through time they had become less functional and more decorative, such that by period 63, they were simply adornments (Fig. 3.8):

> The most clear series of derived forms is that of the wavy handled vases.... Beginning almost globular, with pronounced ledge-handles, waved (as in stage 35 to 42), they next become more upright, then narrower with degraded handles, then the handle becomes a mere wavy line, and lastly an upright cylinder with an arched pattern or a mere cord line around it. (Petrie, 1901:5)

This is phyletic seriation. Once he had the vase sequence worked out, it then became a matter of ordering grave lots based on the vase forms associated with them. Further, vessel types that co-occurred with particular handled vase types then became marker types in their own right and could be used to place correctly other grave lots that did not contain handled vases. This was cross dating, the topic of Chapter 6.

After arranging the vases in sequence, Petrie noted that the contents of the vases changed with vessel form:

> at first full of a strongly aromatic ointment, later with a layer of clay over it, next with mainly clay only scented with ointment, lastly filled with merely solid clay, as in the cylinder jars. The degradation of contents to a worthless substitute proves from which end of the scale the changes proceed. (Petrie, 1901:5)

This orderly change in vessel content suggested to Petrie that his ordering based on change in vessel form and handle shape was correct. Petrie's comment regarding "degradation of contents" shows that he made an a priori assumption about which way time ran, that is, the aromatic ointment was early and the "worthless" substitute was late. Had he actually used that proposed sequence to order the vessels as opposed to it comprising a post facto observation, it would have been an evolutionary seriation.

Petrie (1899a:300) referred to portions of his temporal sequence of pottery forms as "genealogies." Nowhere did he expand this notion, which is unfortunate because of its evolutionary aspects. For example, recalling that low numbers fall early in the sequence and high numbers late, as shown in Fig. 3.9, the jar form in period 38 appears to be a hybrid of two different forms, as does the form in period 70. If a correct interpretation, it illustrates that the evolution of the cultural traits of vessel form is, as Kroeber observed decades later, polyphyletic. But we do not

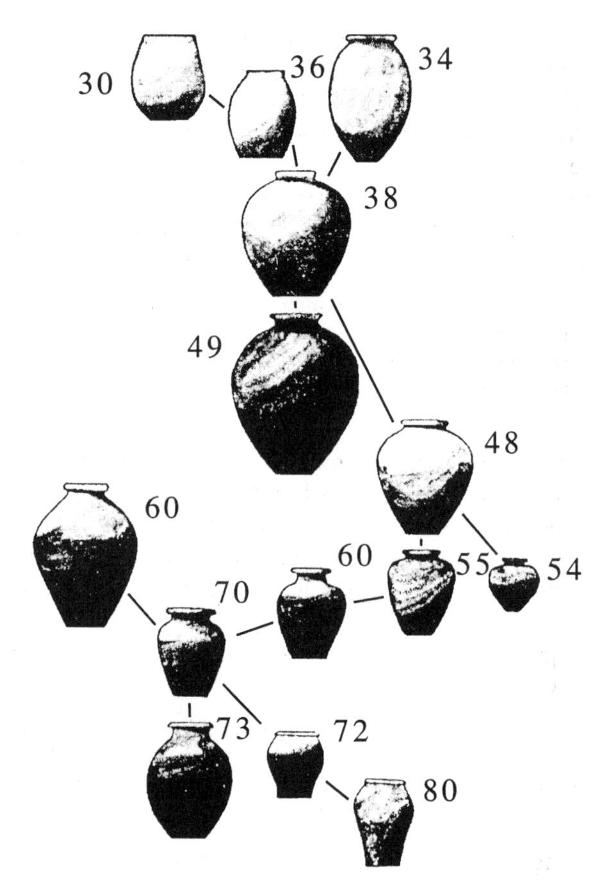

Figure 3.9. W. M. Flinders Petrie's genealogy of ceramic vessel forms recovered from burials in three localities in Egypt. The numbers refer to periods (see Fig. 3.8) (after Petrie, 1899a).

know if this is a correct interpretation, because Petrie made no effort to distinguish between homologous and analogous attributes or between synapomorphies and symplesiomorphies. We should not expect him to have attempted the latter distinction, as such was not done in biology until the 1960s. Why he did not attempt the former reflects, we suspect, the haphazard borrowing of evolutionary concepts from biology typical of archaeologists.

Most interesting of all from the perspective of archaeological dating, Petrie noted in the figure caption that accompanied Fig. 3.9, "These forms pass through two or three different fabrics, showing that form is more important than material" (Petrie, 1899a:300–301). This is an excellent observation that parallels our discussion in Chapter 2 of the plain paste types Neeley's Ferry Plain and Bell Plain and the fact that they are not very good chronological markers. Of much more utility are such things as vessel form and decoration, both of which tend to be more chronologically sensitive than do functional dimensions such as paste.

Once the grave lots were arranged in a temporal sequence based strictly on his arrangement of pottery forms, Petrie noticed that other classes of artifacts—stone vessels, slate palettes, and flint implements—served as checks on the ordering. Figure 3.10 illustrates the ordering of stone vessels as well as Petrie's ideas of how

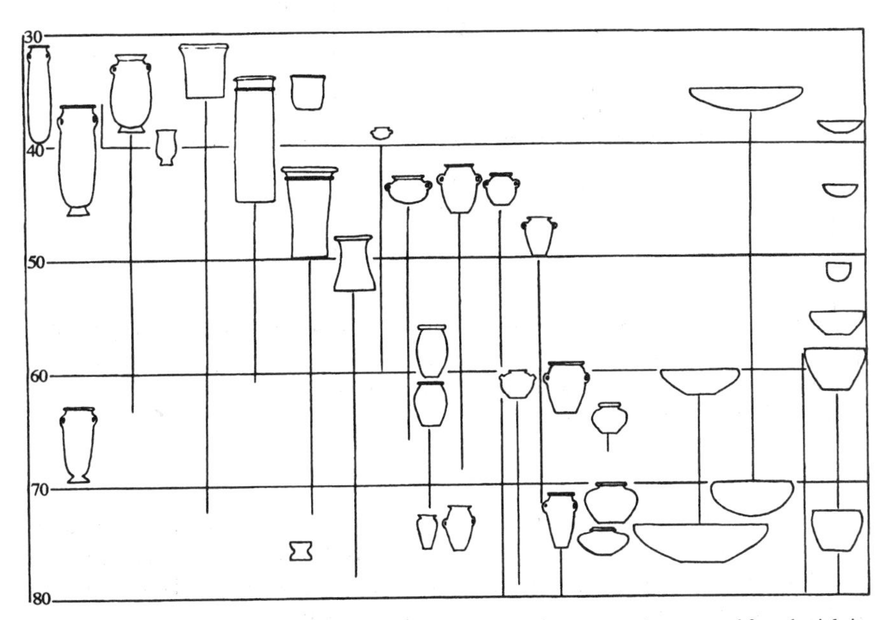

Figure 3.10. W. M. Flinders Petrie's chronological ordering of stone vessels recovered from burials in three localities in Egypt. Periods from 30 to 80 are shown at the left; vessels from each period are placed in rows (after Petrie, 1901).

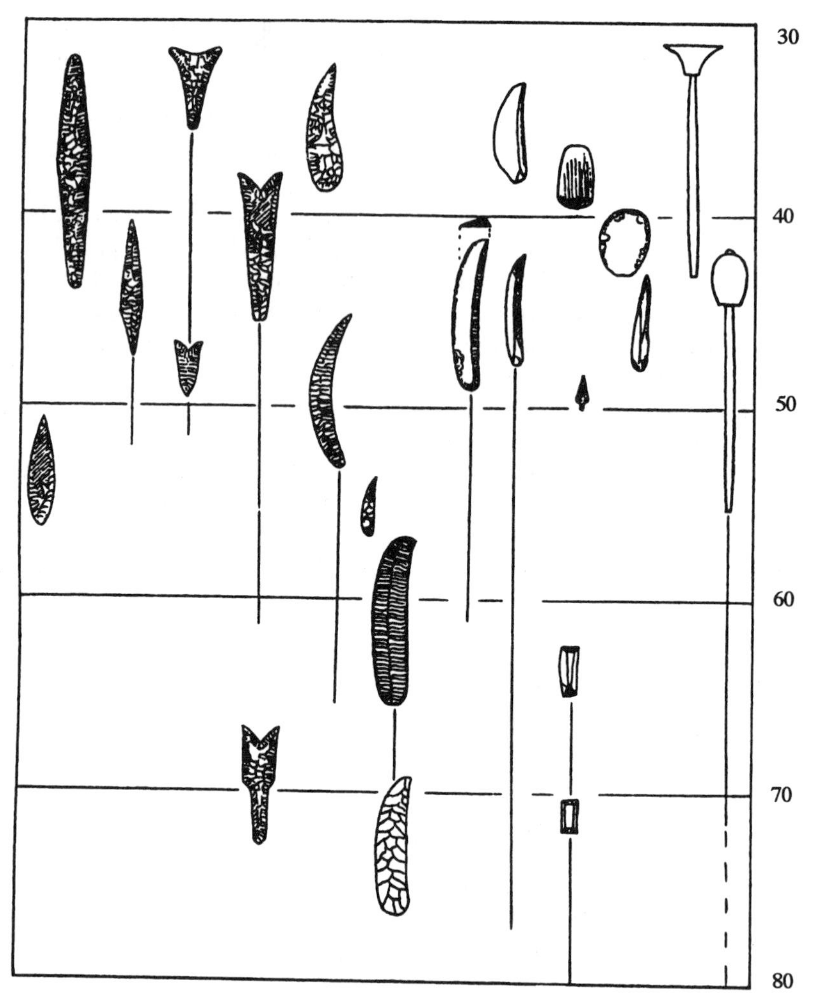

Figure 3.11. W. M. Flinders Petrie's chronological ordering of flint artifacts recovered from burials in three localities in Egypt. Periods from 30 to 80 are shown at the right; artifact forms from each period are placed in rows (after Petrie, 1901).

some of the forms were related phyletically. Figure 3.11 illustrates the chronological positioning of various kinds of flint implements. Petrie created his stone vessel and flint implement types after he knew the correct sequence based on associated ceramic vessels, but they are historical types nonetheless. That is, they have continuous life spans as opposed to disappearing and then reappearing at a later date. Their similarities suggest historical continuity, and thus the passage of time, an inference in need of testing. The phyletic seriations included in Figs. 3.9–3.11 comprise inferences of heritable continuity that also require testing.

John Evans and Gold Coins from Britain

Petrie's use of phyletic seriation as the basis of chronological ordering had precedence in the work of several British archaeologists, including A. L.-F. Pitt-Rivers (1870), who seriated copper and bronze axes, and John Evans (1850), who seriated gold coins from Great Britain that were minted prior to and after the Roman invasion of Britain in 54 BC. Only Pitt-Rivers (1875) was explicit about why heritable continuity provided an explanation for the phyletic seriations of Evans and Petrie. Pitt-Rivers (1875:294, 295) "selected and arranged [artifacts] in sequence, so as to trace, as far as practicable, the succession of ideas," and thus his arrangement illustrated "the development of specific ideas and their transmission from one people to another, or from one locality to another." But whereas Pitt-Rivers was exceptionally explicit about his explanatory theory for both historical and heritable continuity, his predecessors and successors were not.

Evans (1850) used changes in two characters or dimensions of variation to seriate the coins: weight and design. A third dimension, die size, did not produce particularly useful results. Not visible in Evans's seriation (Fig. 3.12) is the decrease through time in coin weight. For example, type 2 coins on average weighed 103.5 grains, type 3 coins 91.5 grains, and type 4 coins 87.25 grains. Highly visible, though, is the change in design on both sides of the coins. The sequence begins with the natural-looking laureated bust of Phillip II of Macedon on the obverse and a horse-drawn chariot on the reverse. Through time, the designs on both sides became successively more stylized until a point was reached at which they again became naturalistic. Evans perceptively identified typological "creep," the problem noted by Phillips, Ford, and Griffin (1951) as occurring when one moves away from the "centers" that produced the analytical prototype:

> Thus far I may observe at present, that the coins generally recede farther from the prototype as the places of their discovery recede from the southern coast—as, for instance, the Yorkshire and Norfolk types Nos. 24 and 16; and that in the southwestern counties the workmanship of the coins appears continually to have deteriorated; while in the southeastern and eastern, after declining for a time, it again improves, probably through the introduction of foreign artists, till, under Cunobeline, it attains its highest perfection. (Evans, 1850:137)

Recall that there are two kinds of similarity—analogous and homologous—and that the latter comprises two sorts—synapomorphic and symplesiomorphic—but only synapomorphies are useful for ascertaining phylogenetic history. Petrie, Evans, and Pitt-Rivers did not make these distinctions explicit. Rather, their work was founded in and originated with the use of the comparative method in linguistic studies of the late eighteenth and early nineteenth centuries (Leaf, 1979:86–90). In

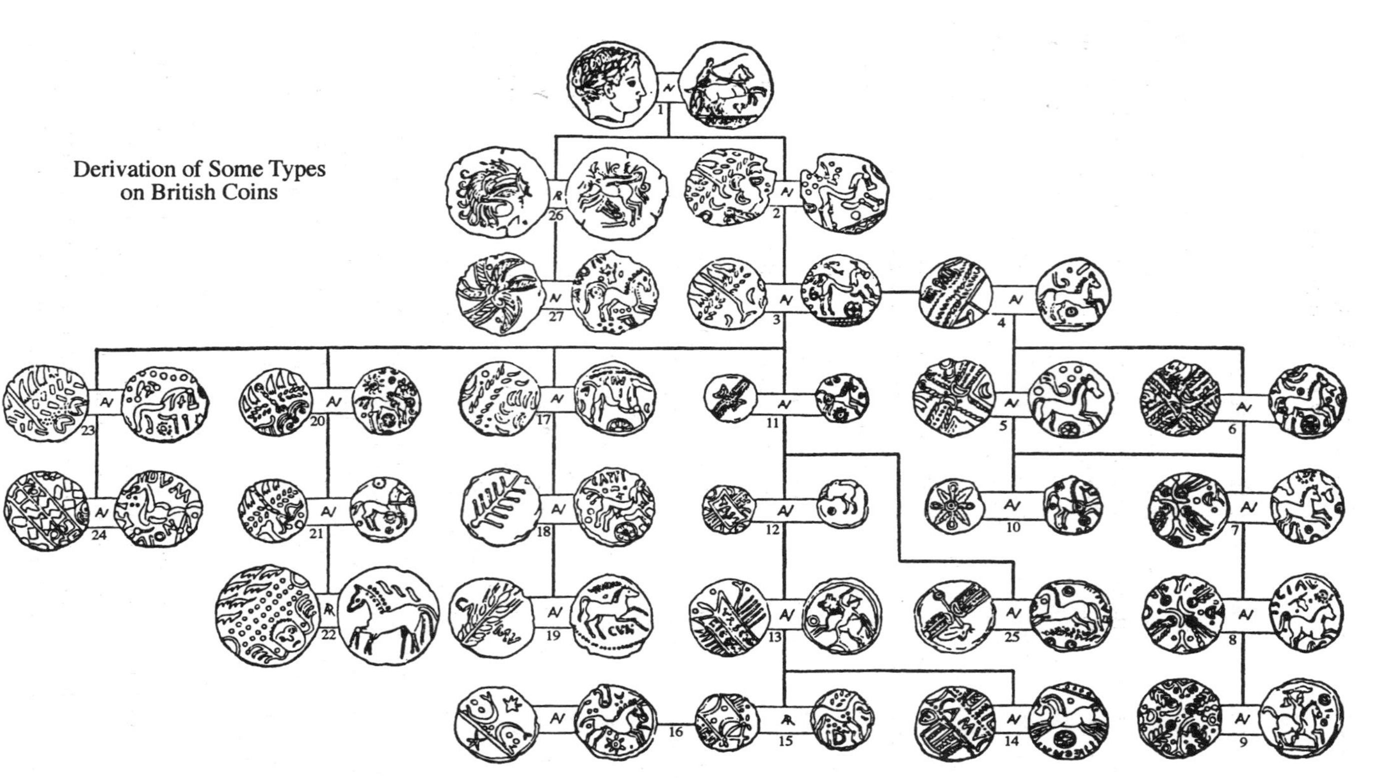

Figure 3.12. Stylistic changes in British coins proposed by John Evans. The sequence begins with the natural-looking laureated bust of Phillip II of Macedon on the obverse and a horse-drawn chariot on the reverse. Through time, the designs on both sides became successively more stylized until a point was reached at which they again became naturalistic (after Evans, 1850).

short, similarity in form denoted historical (and, presumably, heritable) continuity, whether the objects compared were words or artifacts. Such an explanation for similar organisms was not used by biologists until the publication of Darwin's *On the Origin of Species*. Darwin's insight provided a new and logical causal explanation as to why there were formal similarities between organisms and why the Linnaean taxonomy was hierarchical; it reflected descent with modification. Darwin (1859:206) argued that "By unity of type is meant that fundamental agreement in structure, which we see in organic beings of the same class, and which is quite independent of their habits of life. On my theory, unity of type is explained by unity of descent." In two short sentences, Darwin clearly distinguished between analogous and homologous characters. Our point is that the notion that formal similarity denotes evolutionary relatedness, a combination of the assumptions of historical continuity and heritable continuity, has a deep history in the social sciences, deeper than Darwinism and deeper than genetics. In short, an interpretive algorithm for formal similarity founded in common sense was available to archaeologists from the beginning. Only Pitt-Rivers (1875) offered an explicit rationale for why such should hold for cultural phenomena: heritable continuity, and thus historical continuity, especially as indicated by phyletic seriations, reflected empirically the transmission of ideas.

Although used since at least the middle of the nineteenth century by archaeologists in Europe and since late in the nineteenth century in North America, the interpretive algorithm was stated explicitly in the middle of the twentieth century by Gordon Willey (1953:363), who characterized as an unequivocal methodological axiom of culture history the notion that "typological similarity is an indication of cultural relatedness." Nowhere in Willey's discussion is the distinction between analogous and homologous characters made explicit, nor are what we have termed historical continuity and heritable continuity distinguished. Further, by 1953 archaeological "types" generally comprised discrete objects, not their attributes or attribute states (the problem of scale). The general failure of archaeologists to distinguish between analogous and homologous similarities concerned Irving Rouse (1955), who pointed out that there were three steps to determining the historical relatedness of archaeological units. First, determine the extent or degree of their typological similarity. Second, determine their degree of proximity in time and space; contiguity in both denotes the potential for contact or interaction, and thus the potential for an evolutionary relation. To determine if contact had taken place required the third step, which comprised the distinction between analogous and homologous similarities (heritable continuity). From this third step, one could determine the phylogenetic history of the units. But Rouse, like Julian Steward (1929) before him, was not explicit about how the third step was to be accomplished. Within the discipline generally, this role was filled by the axiom explicated by Willey (1953). The result, of course, was and still is endless debate over whether similar archaeological phenomena owe their similarity to common heritage or to adaptive convergence.

Do not misinterpret the preceding statement. It does not mean early Americanist culture historians were ignorant of such problems, as attested particularly by Steward's (1929) and Kroeber's (1931a) efforts. Rather, because most of them built chronologies in areas of restricted spatial extent, the criterion of geographic propinquity was met, as was the criterion of historical continuity when artifacts were recovered from a single column of vertically superposed sediments. The simultaneous if implicit use of spatiotemporal propinquity increased the probability that the phyletic seriations produced were founded on homologous similarity, and thus comprised heritable continuity, though they did not guarantee such. In fact, such similarity was of minor concern relative to the more immediate problem of constructing an ordering of artifact forms that was chronological. It was generally after the ordering was confirmed to comprise a chronological sequence that inferences of whether similarities were homologues or analogues were rendered.

A. V. Kidder and Pottery from Pecos Pueblo

Sixteen years after his excavations at Pecos Pueblo in New Mexico, Kidder claimed he had, prior to those excavations, "attempted a seriation, on comparative grounds, of the material available" (1931:7). Kidder's phyletic seriation technique, probably learned from George Reisner who had worked in Egypt (Browman and Givens, 1996:86) and undoubtedly was familiar with Petrie's work, differed markedly from the frequency seriations of Kroeber and Spier (see Chapter 4). Kidder's (1915) original pottery sequence was based on his suspicions regarding the evolution of various design and technological attributes. In a brief paper published in 1917, he demonstrated how such an evolutionary, and thus temporal sequence, could be worked out. Importantly, he cautioned that the "only safe method for the working out of developments in decorative art is to build up one's sequences from chronologically sequent material, and so let one's theories form themselves from the sequences" (Kidder, 1917:369). The chronological sequence of attributes or character states was confirmed in 1917 by study of superposed collections. The suspected order had been constructed 2 years earlier (Kidder, 1915) using phyletic seriation of the character states of pottery decoration. Figure 3.13 illustrates an example of Kidder's sequence of pottery designs from earliest (1) to latest (5). He proposed that through time, the pottery design became less intricate, changing from a stepped motif (1), to a pair of stepped motifs (2), and finally to a series of ever-larger white spaces (3–5). Kidder here bypassed the notion of type (though it was incorporated into his later work at Pecos) and concentrated strictly on a single character, using changes in character state to tell time.

As brilliant as Kidder's analysis was, few other Americanists tried it, probably because by 1920 superposition and stratigraphic observation had become the technique of choice for creating a chronological order rather than merely a tech-

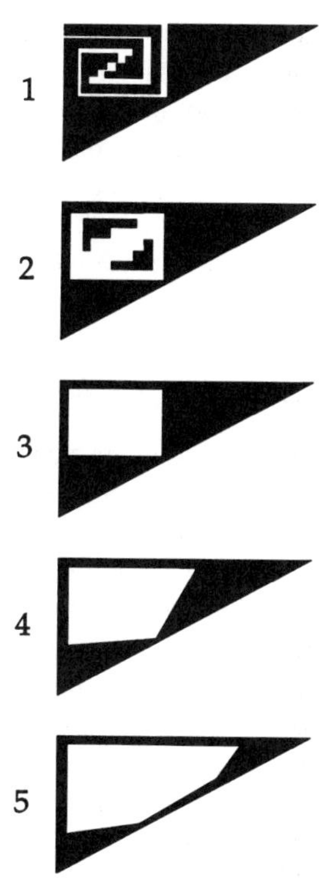

Figure 3.13. A. V. Kidder's illustration of the evolution of a ceramic design on pottery from Pecos Pueblo, New Mexico. The designs are numbered chronologically, with "1" being the earliest. Kidder proposed that through time, the pottery design became less intricate, changing from a stepped motif (1), to a pair of stepped motifs (2), and finally to a series of ever-larger white spaces (3–5) (after Kidder, 1917).

nique for confirming that a particular ordering was in fact chronological (Lyman and O'Brien, 1999). We take up this point in more detail in Chapter 5. Here it suffices to note that Kidder (1917:370) termed the results of his phyletic seriation a "series" and his discussion here and elsewhere (e.g., Kidder and Kidder, 1917) imply that, for him, a series was not only a chronological sequence (denoting historical continuity) but an evolutionary one (denoting heritable continuity) as well. On the one hand, the former was an inference that could be tested with superposition, and Kidder (1916, 1924, 1936a) performed such tests. The latter, on the other hand, was an interpretation that could either be derived after a chronological sequence had been built, or it could be used, as Kidder did, to construct a chronological sequence. Kidder was not explicit about this, nor were his contemporaries or intellectual followers. The result was disastrous, as chronological method and interpretive algorithm were conflated. Sometimes the confirmed chronology (historical continuity) was used in turn to confirm phylogenetic inferences (heritable continuity); other times the suspected phylogeny was the basis of the chronological inferences. Failure to keep the two operations—chronology building and phylogenetic inference—separate resulted from the commonsensical approach to the archaeological record adopted by culture historians (Lyman *et al.*, 1997a,b). A stark case in point is found in the efforts of several archaeologists in the Southwest.

The Gladwin–Colton–Hargrave System

Americanist prehistorians working in the Southwest in the 1920s and 1930s spent considerable effort attempting to standardize pottery descriptions (e.g., Gladwin and Gladwin, 1928, 1930, 1934; Haury, 1936a). It was generally argued that pottery types were supposed to integrate the dimensions of time, space, and form, but they were also supposed to do something else. Types were supposed to reveal phylogenetic connections among Southwestern groups. In naming a type, temporal placement and comparative terms were omitted, and a binomial nomenclature was used. Color or surface treatment constituted the first ("genus") part of the type name, and geographic locale constituted the second ("species") part (Gladwin and Gladwin, 1930). This nomenclature system, which resulted in the " 'family tree' concept of culture classification" (Brew, 1946:58), was in large part the result of the work of Harold Gladwin (e.g., Gladwin and Gladwin, 1930, 1934), Lyndon Hargrave (1932), and Harold Colton (e.g., 1932). Colton and Hargrave's (1937) study of the pottery of northern Arizona is the most revealing, and we focus on it here.

Colton and Hargrave's typological system (Fig. 3.14) had elements of modern dimensional, or paradigmatic, classification (Dunnell, 1971), wherein classes are constructed by linking character states under each dimension. Such a system is ideal for identifying variation and tracking it through time, but in Colton and Hargrave's grand scheme it was simply a means of facilitating communication among archaeologists; it had no scientific work to do. The lack of a scientific reason for creating the classification caused Colton and Hargrave to resort to common sense as their warrant for selecting the particular characters and character states they chose to examine (such as angle of side wall and lip direction). In commenting on the classification, Paul Reiter (1938:490) saw it as "stimulat[ing] suspicion that the potters of northern Arizona must have been parsimoniously impeccable in their adherence to ceramic creeds." Neither he nor Colton and Hargrave saw the originality in the system: using ideational units constructed at the scale of attributes of discrete objects to measure variation in pottery forms. Their failure was a direct result of viewing types of discrete objects as empirical rather than theoretical units.

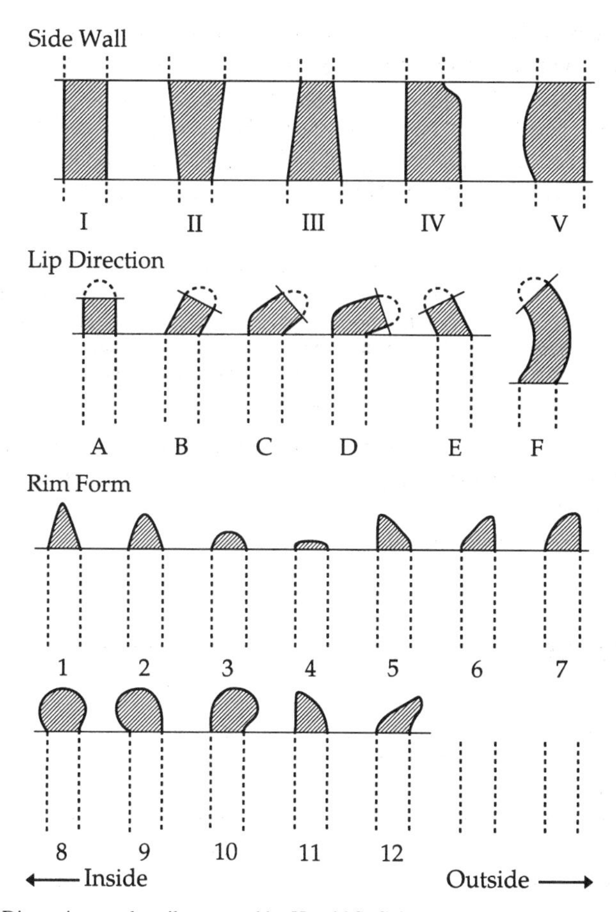

Figure 3.14. Dimensions and attributes used by Harold S. Colton and Lyndon L. Hargrave to classify Southwestern rim sherds. Although they did not use the terms "dimension" and "attribute," their system is a paradigmatic classification (Dunnell, 1971), in which classes are created by the intersection of attributes of analytical dimensions (after Colton and Hargrave, 1937).

Colton and Hargrave were clear about what they thought regarding the historical relatedness of the types they and others created. Their emphasis was on the relatedness of the types of pottery and not on the relatedness between and among the character states of the pottery. This was unfortunate because their classification system was capable of tracking the histories of character states, and thus was not only useful for measuring time's passage but also for distinguishing among analogues, synapomorphies, and symplesiomorphies. They did none of this, however. Colton and Hargrave (1937:3) were more explicit than Kidder had been when they defined a ceramic "series" as "a group of pottery types within a single ware in which each type bears a genetic relation to each other." For them, a ware was "a group of pottery types which has a majority of characteristics in common but that differ in others" (Colton and Hargrave, 1937:2). The relevance of the concept of series for archaeological dating in general and phyletic seriation in particular is clear in Colton and Hargrave's (1937:3) remark that a series should include

> all those types and only those types that occur: (a) in the direct line of chronological genetic development from an original primitive or ancestral type to a late type; and (b) as collateral developments or variations from any type in that line of development, but which are not themselves followed in chronological genetic sequence by derived types other than types derived through the main line of development from the type of which the collateral type is a development or variation.

Their graph showing these relations is reproduced in Fig. 3.15. Although Colton and Hargrave's explanatory theory sounded like biological evolution, they failed to make explicit what that theory comprised, and they did not make clear how it was to be implemented archaeologically.

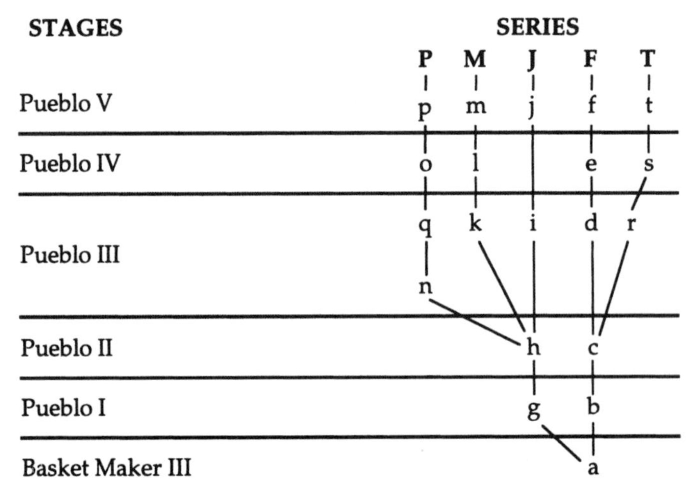

Figure 3.15. Harold S. Colton and Lyndon L. Hargrave's hypothetical representation of the relation between a series and its types. In the example, series F comprises types a–f. Type a is ancestral to all other types; type c is ancestral to types d, e, f, r, s, and t; type c is collateral to types d and r, derivative from type b, and ancestral to types f and t; all types are related to each other through type a; types i, j, k, l, and m, are related to each other through type h; types q and n are both ancestral to type o but collateral to each other and are derivative from types h, g, and a. Note one peculiarity of the system: Colton and Hargrave state that type c is collateral to types d and r, and they also state that types q and n are collateral to each other. This is nonsense, being instead a function of where one draws the stage lines (from Colton and Hargrave, 1937).

In his review of their handbook, Ford (1940:264) lamented that Colton and Hargrave did not consider "the utility of the types for discovering cultural history," though he was fairly optimistic that the genetic relations among types and wares within a series could be determined. He noted, however, that Colton and Hargrave ignored the problem of selecting "a class of features which will best reflect cultural influences [e.g., transmission via contact], and which in their various forms will be mutually exclusive, to serve as guides in the process" of determining ancestral–descendant relationships (Ford, 1940:265). In short, Colton and Hargrave failed to specify how homologous similarity was to be identified and distinguished from analogous similarity. In his review of Colton and Hargrave's handbook, Reiter (1938:490) noted that he "was unable to find a single instance of proof of [the genetic relationships of pottery types]." Reiter (1938:490) also noted that the typology tended to ignore variation and insisted that "variation tendencies cannot be overlooked if genetic or chronologic emphasis is strong." Reiter's comments were on the mark because Colton and Hargrave's types did ignore variation. Further, their types might have reflected historical continuity, and thus been useful for solving local chronological problems, but Colton and Hargrave wanted to use them to reconstruct historical lineages. This was difficult, given that the types had been constructed without explicit consideration of homologous and analogous features.

Many individuals, including J. O. Brew, were not happy with the Gladwin–Colton–Hargrave scheme of establishing phylogenetic relations. Brew (1946:63) held the then-typical conception of cultural change: "We are dealing with a constant stream of cultural development, not evolutionary in the genetic sense, but still a continuum of human activity." But he was more perceptive than many of his contemporaries:

> We must ever be on guard against that peculiar paradox of anthropology which permits men to "trace" a "complex" of, let us say, physical type, pottery type, and religion over 10,000 miles of terrain and down through 10,000 years of history while in the same breath, or in the next lecture, the same men vigorously defend the theory of continuous change. (Brew 1946:65)

The paradox emanates

> from the belief that the manufactured groups [types] are realistic entities and the lack of realization that they are completely artificial.... Implicit in [the belief] is a faith ... in the existence of a "true" or "correct" classification for all object, cultures, etc., which completely ignores the fact that they are all part of a continuous stream of cultural events. (Brew, 1946:48)

The paradox identified by Brew is believing in real types but viewing change as a continuous stream (Lyman *et al.*, 1997b; O'Brien and Lyman, 1998). Brew recommended more new classifications be constructed, but the kinds of classifications and the kinds of units comprising these classifications that were necessary to resolve the paradox were unspecified.

The most problematic part of the Gladwin–Colton–Hargrave scheme was its evolutionary implications, which Brew (1946:53) found unacceptable: "[P]hylogenetic relationships do not exist between inanimate objects." As he pointed out, biologists were aware that the Linnaean taxonomic system did not necessarily denote phylogeny, though it might. Further, Brew (1946:55) argued that "The only defense there can be for a classification of [artifacts] based upon phylogenetic theory is that the individual objects were made and used by man." Apparently unaware of Pitt-Rivers's (1875) early remarks, Brew did not even accept this notion, because for him evolution involved only the processes of genetic transmission and genetic change. Thus, he perceived only a weak correlation between an organism and the "artifacts" that that organism might produce, such as eggshells and birds or mollusk shells and mollusks, and no connection at all between people and their artifacts. He quoted a single biologist—geneticist Thomas H. Morgan—who argued that a phylogenetic history did not explain organisms; hence to Brew it could hardly explain artifacts: "This is a most important point, and I wish to emphasize it here" (Brew, 1946:56).

Despite Brew's concise arguments, the axiom "typological similarity is an indication of cultural relatedness" (Willey, 1953:363) survived unscathed, whereas the Gladwin–Colton–Hargrave system fell into disuse. Gordon Willey and Philip Phillips (1958:31), for example, argued that inferences of "cultural relatedness" were desired and that such inferences demanded interpretive concepts that were "culturally determined." They suggested one must identify traditions, horizons, and horizon styles, because these notions were founded in ethnological reality and denoted "some form of historical contact" rather than "phylogeny" (Willey and Phillips, 1958:30). Traditions, originally defined as "a line, or a number of lines, of [artifact] development through time within the confines of a certain technique or decorative constant" (Willey, 1945:53), denoted enculturation and persistence or transmission over time; that the definition and implications of the term tradition were the same as those for a biological lineage went unremarked. Horizons and horizon styles, originally explicitly defined as "the recurrence of specific features of style or manufacture in prehistoric artifacts ... from one region to another so that the phenomena [allow us to] coordinate our knowledge of the past in a broad temporal and spatial scheme" (Willey, 1945:8), denoted diffusion or transmission over space (we return to horizons and traditions in Chapter 6).

To Willey and Phillips and most other culture historians of the 1950s, "phylogeny" denoted a genetic connection, abhorred since Brew's (1946) scathing critique of the Gladwin–Colton–Hargrave system. Thus Robert Ascher (1963:571), in reviewing Ford's (1962) manual for building cultural chronologies, noted that Ford referred "to sherds [using] such terms as 'descendants,' 'ancestral forms,' and 'parallel lineages.' Sherds were never alive: is it not time to drop denotative diction derived from questionable analogies?" But the contradiction internal to Willey and Phillips's position, held as well by most of their contemporaries, escaped notice. How could historical contact between cultural or artifact lineages or continuity within a lineage not be phylogenetic, albeit nongenetic, processes? How can one subscribe to the notion of historical continuity and use it as a principle for ordering phenomena chronologically and simultaneously discard the notions of heritable continuity and phylogeny? Simply put, one cannot; heritable continuity is, by definition, phylogenetic, regardless of the mechanisms of transmission and regardless of what is transmitted.

PROJECTILE POINT EVOLUTION

If the phenomena under study evolve, then knowing the phylogenetic history of those phenomena is in many respects a required part of the explanation of those phenomena. This is as true in archaeology as it is in biology (Lyman and O'Brien, 1997, 1998; O'Brien and Lyman, n.d.; O'Brien *et al.*, 1998). How can we explain the evolution of a species if we cannot organize the character states of individuals of that species into a chronological sequence? We need to know when certain character states make their initial appearance, when they disappear, and when one state replaces another. And, for our purposes here, such an ordering gives us an excellent set of chronological tools. Regardless of whether one buys into evolutionary archaeology and its emphasis on artifacts as phenotypic expressions, it is difficult to deny that the historically oriented methods advocated by evolutionary archaeologists have significant chronological implications. Those methods bypass the problems associated with earlier evolutionary efforts in archaeology (e.g., Lyman and O'Brien, 1998), such as the Gladwin–Colton–Hargrave system—because they emphasize simple ordering based on homologous similarity and explicitly recognize the assumptions of historical continuity and heritable continuity.

As an example of how one might approach a phylogenetic analysis in archaeology, we focus on the evolution of a kind of projectile point found over much of the midwestern and southeastern United States: the Dalton point, easily recognized by its distinctive lanceolate shape, characteristic parallel flaking, concave base, and often beveled blade (Fig. 3.16). The question is, when did Dalton points first appear and how long did they last? Had archaeologists taken seriously Kidder's phyletic seriation technique in combination with Steward's and Kroeber's admonitions, there would have been no long-lasting debates about the proper chronological position of Dalton. But this was not the case. Dalton as a type of projectile point became an empirical unit and in the process archaeologists lost any ability to examine the tremendous variation exhibited by specimens placed in that type. That variation, especially when viewed against variation in several other projectile point types, tells us exactly where Dalton points fall chronologically.

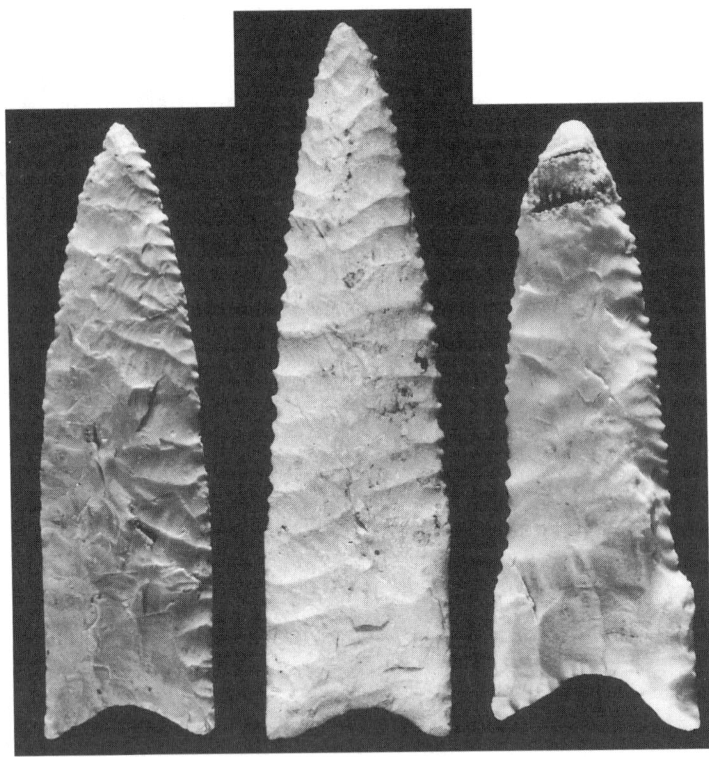

Figure 3.16. Dalton points from eastern Missouri. Some differences in blade shape reflect amount of resharpening, but some undoubtedly reflect differences in the way the points were designed to be used. Despite the differences, all the points would be placed in the Dalton type. Specimen second from left is 9.6 cm long (from O'Brien and Wood, 1998).

The earliest points found in the Midwest and Southeast appear to be Clovis points (Fig. 2.7), which based on uncalibrated radiocarbon dates from contexts in the Plains and Southwest fall within a narrow range of time, perhaps 9250 BC to 8950 BC. The dating of Clovis points in the Midwest and Southeast is more problematic, because few radiocarbon dates exist, and the ones that do suggest that Clovis points in those regions might be younger or older than those to the west. A host of radiocarbon dates from numerous midwestern and southeastern sites (O'Brien and Wood, 1998) indicate that Dalton points postdate Clovis points, and most archaeologists assumed that Dalton point manufacture might have begun around 8000 BC and ended sometime around 6500 BC. We, like Albert Goodyear (1982), were skeptical of many of the late Dalton dates (some extending up as late as 5000 BC) reported from various sites in the Midwest and Southeast. However,

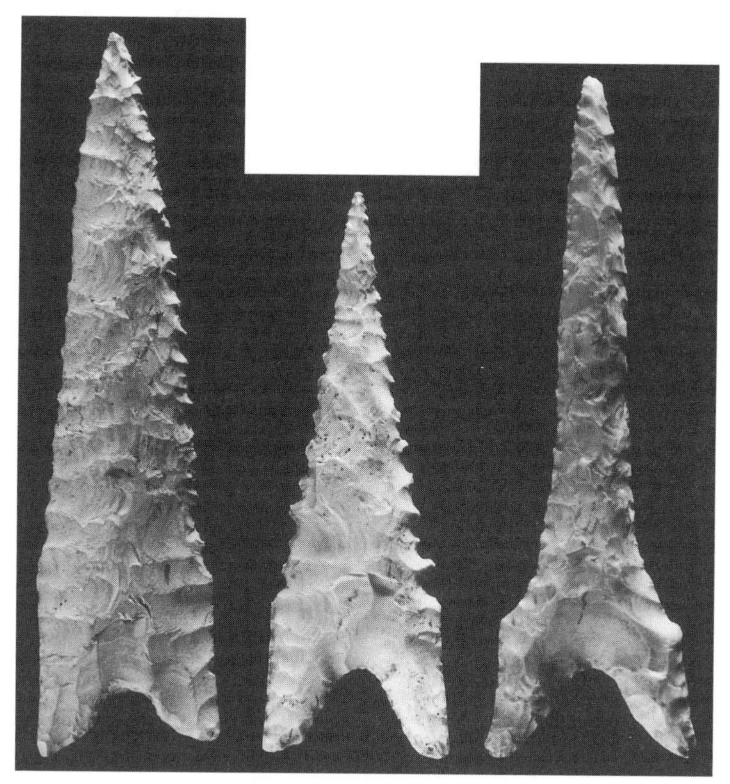

Figure 3.16. (*Continued*)

we thought that Goodyear was too conservative in his estimate that Dalton points ceased to be manufactured around 7900 BC, especially in light of the fact that the beginning date was at best a few hundred years earlier. We have since changed our minds (O'Brien and Wood, 1998). The one site in the United States that has produced what is in our minds (as it was in Goodyear's) the only acceptable radiocarbon dates (8580 ± 650 BC and 8250 ± 330 BC) for Dalton points is the terrace in front of Rodgers Shelter in southwest Missouri (O'Brien and Wood, 1998).

It now is obvious that Goodyear was correct: Dalton points date no later than about 7900 BC. It also is obvious that Dalton points began to be manufactured by about 8900 BC, right on the heels of Clovis points, thus for a time becoming the midwestern and southeastern temporal equivalent of the Folsom point, which typically is restricted to Plains states and dates roughly 8950–8650 BC. We base our opinion of the beginning date of Dalton points not on any radiometrically dated contexts—the dates for Rodgers Shelter are not that early (see above)—but rather on the apparent lack of points in many areas of the Midwest that date to the ninth millennium BC. It does not seem reasonable that Clovis points just happened to last much longer in Missouri than they did on the Plains and in the Southwest, nor is it reasonable to assume that people stopped making Clovis points and then waited around for a millennium or so before they made points again. The answer to what occurred between Clovis and Dalton has been staring us in the face for years, but until recently (O'Brien, 1998; O'Brien and Wood, 1998) we never saw it. The reason we never saw it is rooted in how we had come to view the archaeological record. We had become so attached to the units we use to categorize artifacts—the units termed Clovis, Folsom, Dalton, and the like—that they began to take on a reality of their own. Then, based on the clear replacement chronologically of Clovis points by Folsom points on the Plains and in the Southwest, we became convinced that that same replacement must have occurred elsewhere as well. The fact that few Folsom points have been found east of the Plains was simply written off as a sampling problem.

If archaeologists had originally paid more attention to variation, they might quickly have seen where Dalton points fell chronologically. They certainly exhibit characteristics in common with numerous early points, including those placed in the Clovis type. On the whole, Dalton points are thinner than Clovis points, though flaking patterns are similar. Most people can distinguish the two, but there are numerous specimens that are difficult to place in one type or the other. For example, take a close look at the point found near Independence, Arkansas, pictured in Fig. 3.17. This 13-centimeter-long specimen is less than half a centimeter thick at its thickest point, thinner than the majority of Clovis points but well within the range of "typical" Dalton points. The specimen also has the deeply concave base found on Dalton points, but it has fluted surfaces like those on Clovis points. What does one do with such a specimen? One of two rather obvious options has usually been chosen: Create a new type, or place it in one of the two existing types. If the second option is selected, one could add a footnote stating that the point exhibits features contained in the descriptions of both types. In practice, it probably does not matter too much which option one chooses.

Of more importance is recognizing the point for what it is: a form morphologically between Clovis and Dalton that probably dates around 8950 BC or so. It is these "transitional" forms that too often are overlooked by archaeologists intent on placing a particular specimen into one type or another. But such forms are very significant both from the perspective of chronology and for what they tell us about subtle technological changes that were occurring over time. What we normally see pictured in projectile point type guides are "archetypal" specimens: those examples that supposedly are typical of what is contained in the type description. We might talk about variation among, say, Clovis points, but in our

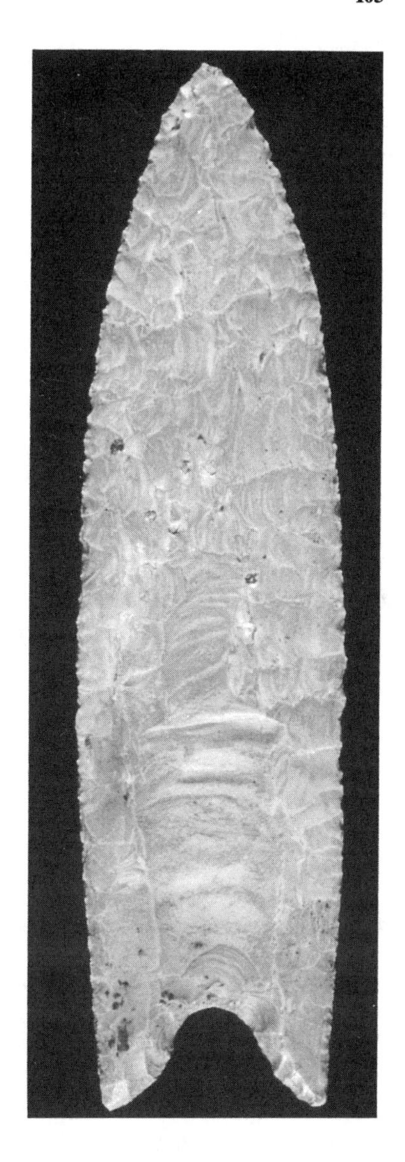

Figure 3.17. Projectile point (13.0 cm long) from Independence County, Arkansas, showing characteristics of specimens placed in either the Clovis or Dalton type. It is such "hybrid"-type specimens that are so valuable in phyletic seriation (from O'Brien and Wood, 1998).

minds there is an archetype of what a Clovis is supposed to look like, and it is that mental picture that guides what we do with points that begin to fall on the margins of the description of the archetype.

The important thing is that the example demonstrates the "transitional" nature of artifacts. Why would we not expect that if Dalton points fell closely on the heels of Clovis points in parts of the Midwest and Southeast, that Clovis points manufactured around 8950 BC would take on some of the characteristics of archetypical Dalton points? Why, for example, would we not expect to see the bases on fluted points become more concave or for the basal feet to become slightly flared? Based on published criteria of the Dalton type, it is perfectly acceptable for a specimen placed in that type to have a deeply concave base and eared feet; it would seem that the only criterion that would exclude the point illustrated in Fig. 3.17 is that it is fluted. Fluting is a key characteristic of Clovis points, which we know were manufactured before Dalton points. Even if we did not know how much time had elapsed between the end of Clovis point manufacture and the beginning of Dalton point manufacture, specimens such as that in Fig. 3.17 give us a clue as to the age. Such a point is a direct phyletic bridge between Clovis and Dalton. But, because of received wisdom that Folsom followed Clovis, which it did on the Plains and in the Southwest, and because Dalton points in the Midwest and Southeast have been found in what appeared to be the same contexts as presumed later points, we have overlooked the rather obvious fact that Dalton point manufacture fell directly on the heels of Clovis point manufacture.

What we really are saying is that sometime in the ninth millennium BC, people in the Midwest and Southeast (1) quit fluting their points, perhaps because the points were thin enough without the flutes; (2) continued to thin the bases; and (3) in some cases began to chip and/or grind the lateral edges of the haft areas to the point that they created shallow side notches such as those evident on some of the specimens in Figs. 2.7 and 3.16. In other words, Clovis points had "evolved" into Dalton points. Bruce Bradley, an expert in prehistoric flint-knapping techniques, recently echoed our point:

> I believe that the data currently point to an in situ technological development of Dalton points directly out of a Clovis technology. Resemblances [of Dalton points] to post-Clovis Paleoindian styles in the High Plains are superficial and at most represent a common origin out of a Clovis predecessor. (Bradley, 1997:57)

If Bradley and we are correct, then the "superficial" resemblances between Dalton points and more-or-less contemporaneous Plains points comprise symplesiomorphies: homologous similarities that are shared both by Dalton points and their contemporaries in the Plains and by Clovis points. The attributes, or character states, of archetypical Dalton points that distinguish them from archetypical Clovis points are synapomorphies.

We are not speaking metaphorically when we note that Clovis points apparently "evolved" into Dalton points, though perhaps more appropriately we should talk about the evolution of point technology or about the evolution of point-making behaviors among prehistoric peoples. This highlights the apparent phyletic relation—heritable continuity—between points placed in various types, a relation that is apparent in the (inferred) homologous morphological characteristics of what typically are termed Clovis and Dalton points. Once we begin thinking in those terms, our attention becomes focused on often subtle variation in such things as the hafting area of a projectile point. Producing a notch on the edge of a haft element could be accomplished by grinding as well as by chipping. Thus it is not too difficult to see a progression from Dalton points into such things as Hardaway side-notched points (Fig. 3.18). Joffre Coe, who named the Hardaway point (Coe, 1964:67) based on examples from the site of the same name in the Piedmont region of North Carolina, must have thought there was enough similarity that he even created a new type, Hardaway Dalton (Fig. 3.18), to deal with specimens that had broad side indentions as opposed to narrow notches.

What we are doing here is building a sequence of point types using subtle changes in characters to link types that are adjacent chronologically. Although we are employing archaeological types, the importance of this kind of analysis transcends such units; we are using the type names Clovis and Dalton only because they have archaeological prominence. We admit that those two types serve their intended purpose fairly well in that they partition the projectile point continuum into chunks that make sense chronologically, but they make a whole lot more sense

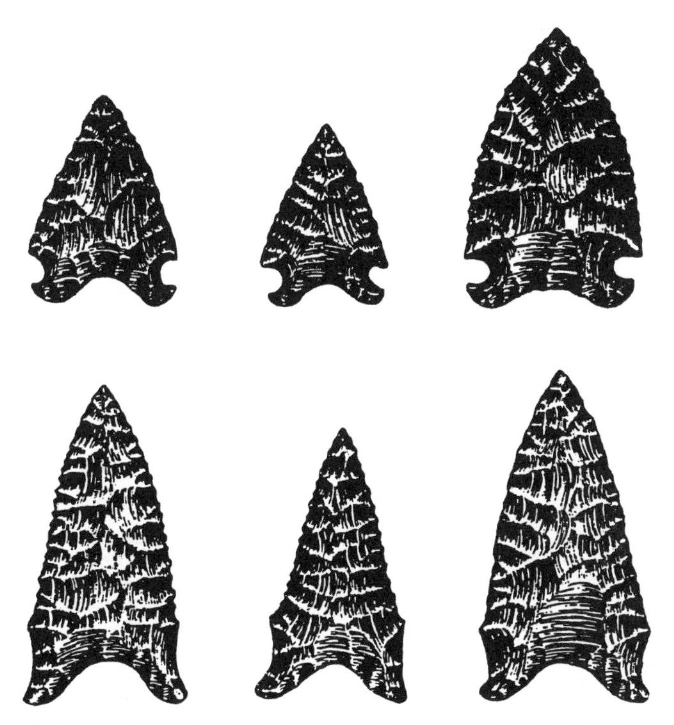

Figure 3.18. Hardaway (top row) and Hardaway Dalton (bottom row) points from the southeastern United States. Specimen in the upper left is 4.8 cm long (after Perino, 1985).

when viewed from the perspective of continuous variation and phylogenetic history. In other words, we need to keep in mind the distinction between theoretical and empirical units. As Phillips *et al.* (1951) warned us, do not confuse categories with things placed in those categories.

In summation, seriation as a dating method rests on the related but distinct assumptions of historical continuity and heritable continuity. When combined, these two assumptions are expressed by archaeologists as historical, or cultural, relatedness and are thought to be manifest as what are termed archaeological traditions. Historical continuity assumes that similar forms will be similar in age. Heritable continuity assumes that the similarity of forms is the result of transmission or heritability. Phyletic seriation, though a relative dating technique not frequently used by late twentieth-century archaeologists (see Deetz and Dethlefsen, 1971, for a notable exception), provides the most straightforward expression of the two assumptions. The assumption of heritable continuity serves as the warrant for interpreting similarity in form as measuring the passage of time, but when used as an interpretive algorithm, it demands that homologous similarities, particularly synapomorphies, be recognized. The seriation method, then, rests on measuring similarities among phenomena. Phyletic seriation focuses on how the characters of a general kind, or type, of phenomena change states over time. As we will see in Chapter 4, there are other ways to measure similarity, and they serve as important checks on the heritable continuity assumed by phyletic seriation.

4

Seriation II

Frequency Seriation and Occurrence Seriation

As should be clear from Chapter 3, the seriation method rests on two assumptions: historical continuity and heritable continuity of form. Heritable continuity has not been explicitly acknowledged in phyletic seriations probably because it is a commonsensical notion that the evolution of artifact forms comprises changes in the characters or attributes of the artifacts. As art historian and archaeologist William Biers (1992:25) observed,

> The particularly characteristic or distinctive way an object appears to the eye can be said to be its style. A change in its appearance, or details of its appearance, or attributes, is seen to be a stylistic change or development.... Whenever we can tell the difference in appearance between two objects of the same type, we are observing stylistic change.

Of course, things are not quite so simple, for as Biers (1992:25) also notes, "Stylistic change can be related to time, but is not necessarily always caused by the passage of time, and can be slow to almost nonexistent, depending on a variety of factors." Yet, the basic notion of stylistic change over time is phyletic: "An evolutionary way of viewing stylistic development is common for ancient art, and is perhaps an influence from the natural world in which biological principles of birth, growth, and death can be observed" (Biers 1992:26).

Phyletic seriations represent historical continuity within a lineage of forms of a kind of artifact such as handled vessels, gold coins, or projectile points. To ensure that the resultant ordering represents and thus measures the passage of time minimally requires an independent measure of time such as finding the same order of artifact forms produced by a phyletic seriation in a column of superposed sediments. This was precisely A. V. Kidder's reason for excavating Pecos Pueblo (Kidder, 1916) and selecting the testing procedure he used there (e.g., Kidder and Kidder, 1917). To ensure that heritable continuity is included in a phyletic seriation requires the identification of homologous similarities, particularly synapomorphies. To date, archaeologists have made serious efforts at meeting the first but not the second requirement. Perhaps this is because few Americanist archaeologists have performed phyletic seriations; most of them opt for frequency (and less often occurrence) seriations where the assumption of heritable continuity, as we will see in this chapter, is built in by the procedure, but such is certainly not explicit in the literature.

As also should be clear from Chapter 3, the process of ordering that comprises the seriation method rests on measuring and evaluating the similarity of the phenomena to be ordered. Phyletic seriations are typically constructed on the basis of the similarity of artifacts at the scale of the attributes they share (Rowe, 1959). The more attributes shared by two objects, the closer they are placed to one another in an ordering and the closer they are thought to have occurred in time. Other ways to measure similarity can be and were conceived by archaeologists early in the twentieth century. Basically, these involve either or both a shift in the conception of units and a shift in scale from the attributes of artifacts to the attributes of aggregates of artifacts. In Chapter 3, we were a bit ambiguous regarding whether a phyletic seriation involved empirical or theoretical units, regardless of scale. It is our distinct impression that the units in such seriations are at least initially treated as empirical—one form blends into another—such as is indicated by Kidder's phyletic seriation of decorative features (Fig. 3.15) and Petrie's phyletic seriations of pottery (Figs. 3.8 and 3.9). These units might eventually become theoretical, but they begin as empirical units. Contributing to the appearance that the units are empirical is the fact that no overlap (see below) is acknowledged, and change thus appears to be transformational, that is, from one form to another.

The shift in scale that occurs with occurrence and frequency seriation, though not mandatory to such seriations as we will see later in this chapter, results in the ordering of aggregates of artifacts variously termed *assemblages*, *components*, *tool kits*, *collections*, or the like; we use the terms assemblage and collection here as these tend to carry fewer connotations of the human behavioral or anthropological significance of the aggregates. Similarity is measured by first noting the presence–absence of types of artifacts or the frequencies of types of artifacts in each aggregate and then either subjectively (visually) or objectively (statistically) ascertaining how similar various aggregates are to one another in terms of the types they share or in terms of the frequencies of the types. In short, the more types shared by two assemblages, the more similar they are, and thus the closer in time the two are thought to be; likewise, the more similar the frequencies of shared types of artifacts in two compared assemblages, the closer to one another the two are placed in an ordering and the closer in time they are thought to be. These comprise occurrence seriation and frequency seriation, respectively. Importantly, because the similarity of assemblages is measured in terms of shared types or type frequencies, the types must be theoretical units, or classes. Only by being such can they have distributions over time and space rather than locations, such as empirical units do. Specimens, or empirical units, have specific locations; classes have distributions that are indicated by the summed locations (in time and space) of the specimens included within them (Dunnell, 1970:307–308).

The occurrence and frequency seriation techniques employ the notions of historical and heritable continuity in somewhat different fashion than what phyletic seriation does. Further, the former two entail certain expectations about how change will appear, not its direction, and thus how time can be measured. We therefore need to outline the various assumptions and expectations that guide the workings of the techniques of occurrence and frequency seriation. Together, these comprise what we refer to as the *seriation model*, which is actually little more than a statement regarding the expected temporal distribution of specimens classified as historical types. We touched on some of these in passing in Chapter 2, and in this chapter we make them explicit and explicate their role within the seriation method. But in order for them to make sense, one must first know at least a little about frequency seriation. Thus we first briefly review the history of frequency seriation before turning to a discussion of the seriation model. Then, we discuss and provide examples of the occurrence and frequency seriation procedures, and consider how one can determine if the requirements and conditions of seriation have been met. We conclude the chapter with a discussion of the temporal resolution provided by seriation, including what has been termed *absolute seriation.*

THE FIRST FREQUENCY SERIATION

Rarely in the history of science is the origin of a major innovation clear. Rather, it is frequently the case that the origin is lost amid different versions of history, and after a time it becomes difficult if not impossible to sort out one version from another. Frequency seriation, however, is another story. A. L. Kroeber gets the credit for this remarkable innovation, and as we have noted elsewhere (Lyman *et al.*, 1997b, 1998) it is clear, despite claims to the contrary (e.g., Browman and Givens, 1996; Praetzellis, 1993; Willey and Sabloff, 1993), that earlier phyletic seriational studies, such as those by Petrie (1899a,b, 1901), Evans (1850), and Uhle (1903), played no role in the development of the technique.

While walking across the countryside around Zuñi Pueblo in New Mexico, Kroeber collected sherds from the surfaces of more than a dozen prehistoric sites. He noticed that some collections tended to be dominated by "red, black, and patterned potsherds," whereas other collections were dominated by white sherds (Kroeber, 1916b:8). He concluded that

> There could be no doubt that here, within a half hour's radius of the largest inhabited pueblo [Zuñi], were prehistoric remains of two types and two periods, as distinct as oil and water. The condition of the sites indicated the black and red ware ruins as the more recent. (Kroeber, 1916b:9)

Based on historical evidence and on the condition of the sites, Kroeber (1916b:9–10) concluded that concerning "the type and period of white ware and the type and period of black and of red ware, the latter is the more recent [belonging] in part to the time of early American history; the former is wholly prehistoric." Thus,

Kroeber had two lines of evidence to indicate that he was dealing with temporal differences in the kinds of pottery he found: (1) historical data, which indicated that at least several of the ruins were inhabited during the period of Spanish occupation, whereas other ruins were said by his informants to have been inhabited long ago; and (2) "Type A [historic] ruins normally include standing walls, and loose rock abounds. All Type B [prehistoric] sites are low or flat, without walls or rock, and … it seems more likely that this condition is due to the decay of age, or to the carrying away of the broken rock to serve as material in the nearby constructions of later ages" (Kroeber, 1916a:43).

Various lines of evidence—historical data, tree rings and the diameters of trees, and stratigraphic positioning—had been used to help measure time prior to Kroeber's work, and the condition or extent of decay had also been used by individuals investigating chronological aspects of the archaeological record. For example, Kidder (1915) used the deterioration of pueblos as an indication of relative age. Differences in pottery associated with the two kinds of ruins—historically documented and not deteriorated, and deteriorated and not historically documented—were thus generally accepted within the discipline as chronological indicators.

Kroeber (1916b:8) "attempted to pick up all sherds visible in certain spots [of each site], rather than range over the whole site and stoop only for the attractive ones." This was the same warning that Ford (1962) later gave. Kroeber did not excavate:

> I have not turned a spadeful of earth in the Zuñi country. But the outlines of a thousand years' civilizational changes which the surface reveals are so clear, that there is no question of the wealth of knowledge that the ground holds for the critical but not over timid excavator. (Kroeber, 1916b:14)

But he was cautious, noting that for his proposed chronological classification, the "final proof is in the spade" (Kroeber, 1916b:20). He also lamented that "in the present chaos of knowledge who can say which of these differences [in frequencies of sherd types] are due to age and which to locality and environment?" (Kroeber, 1916b:21). Kroeber arranged his surface collections to derive not simply a two-period sequence but a six-period (if one includes modern Zuñi pottery) cultural sequence. The innovative aspect of what Kroeber did has been overlooked in Americanist archaeology. Rowe (1962a:400), for example, stated that "Kroeber's seriation of Zuñi sites on the basis of surface collections was not the first successful seriation in North America; A. V. Kidder [1915] had published a seriation of the pottery of the Pajarito Plateau [New Mexico] the previous year." This is true, but Kidder's seriation, as we saw in Chapter 3, was a phyletic one. What Kroeber invented was frequency seriation.

His frequency seriation (Table 4.1) began with corrugated ware as the oldest and most frequent type. Collections of pottery from individual sites were arranged

Table 4.1. A. L. Kroeber's Frequency Seriation of Pottery Sherds from Sites around Zuñi Pueblo, New Mexico[a]

Period	Site	Corrugated	Three color	Black on red	Any red	Black
Present	Zuñi	0	12	1		
Late A	Towwayallanna	1	8	3		
	Kolliwa	—	7	2		
	Shunntekya	2	7	2		
	Wimmayawa	2	4	1	22	53
	Mattsakya	3	4	3		
	Kyakkima	4	3	2		
Early A	Pinnawa	10	1	8		
	Site W	24	—	1		
Late B	Hattsinawa	27	—	5	10	19
	Kyakkima West	12	—	4	8	—
Middle B	Shoptluwwayala	40	—	2	3	7
	Hawwikku B	49	—	6	12	9
Early B	Te'allatashshhanna	66	—	—	—	5
	Site X	71	—	—	3	1
	Tetlnatluwwayala	72	—	—	2	—
Uncertain	He'itli'annanna	—	—	—	—	3
	Site Y	—	—	—	—	—

[a]From Kroeber (1916b).

so that the relative abundance of that type decreased monotonically, with two exceptions (so this is a rather clean solution in the table). The basis for this arrangement was Kroeber's impression that corrugated ware, given its rare association with modern pottery types and its regular association with decayed ruins, decreased in frequency as time passed; this allowed him to arrange the sites in order accordingly. The relative abundance of Kroeber's "three color" type increased monotonically once it appeared in the sequence and was most abundant in the modern Zuñi assemblage. Frequencies of his "black on red" type merely tagged along and fluctuated in abundance, but his "any red" and "black ware" types tended to decrease monotonically. Kroeber initially identified ten pottery types but lumped five of them into two types (and ignored two others) for purposes of his seriation. Ultimately, he used the relative frequencies of only three types to seriate his period A sites and presented only the summed site frequencies for the other two types (lumped from the original five). Kroeber justified this lumping by noting that variations in the frequencies appeared to be a result of sampling error. Such lumping indicates that Kroeber conceived of his types as theoretical units rather than as empirical ones; that is, they were not discovered but were, rather, created by the analyst. They were theoretical units that allowed, if properly constructed, measurement of differences through time.

Kroeber's (1916a:44) chronologically sensitive pottery types indicated that his short periods "shade[d] into one another," and that there was "no gap or marked break between periods A and B." On the one hand, Puebloan culture change was a flowing stream. The two major periods might have originally been "as distinct as oil and water," but they had originally been distinguished on the basis of criteria (e.g., degree of deterioration of associated ruins) other than those used in the seriation. Kroeber (1916b:15) believed that the two major periods "can normally be distinguished without the least uncertainty, and the separateness of the two is fundamental, [but] nevertheless they do not represent two different migrations, nationalities, or waves of culture, but rather a steady and continuous development on the soil." How did Kroeber, then, come up with his early, middle, and late periods? His divisions appear to have been founded on the magnitude of differences in the relative abundance of corrugated ware. Within each period the average difference in relative abundance of corrugated ware between assemblages is 5.2%; the average difference in the relative abundance of corrugated ware between adjacent assemblages assigned to different short periods is 13.5%. The range of differences between assemblages within the same period is 0–15%, and the range between adjacent assemblages of different periods is 3–28%. Thus Kroeber may have maximized within-group homogeneity and between-group heterogeneity. Otherwise, his periods are arbitrary, which is not to say that they are poor constructs. As we will see, such arbitrarily defined periods are typical of seriations and in many cases are to be preferred.

HOW DO OCCURRENCE AND FREQUENCY SERIATION WORK?

For any seriation technique—phyletic, occurrence, frequency—to work, the rate of change within a lineage must be gradual. Although such cannot be assumed literally, it is precisely such an assumption that allows the positioning of like adjacent to like and unlike to be placed some distance apart in an ordering. But do not be misled by the word "gradual." Characterizing the rate of change as "gradual" does not preclude rapid change or fluctuation in the rate of change during the history of a lineage, nor is stasis precluded, but rather only abrupt or sudden change of great magnitude is precluded. In other words, by assuming change is "gradual," one is saying that time can be measured as a continuum rather than as discontinuous chunks. As George Cowgill (1972:384) makes clear, rather than assume change is gradual and continuous

> for seriation to be useful as a basis for chronology, it seems to me that all that is required is that there never, among the set of units being seriated, be a break in the sequence so

> abrupt and catastrophic that units immediately following the break bear no (or only accidental) resemblance to units before the break.

Cowgill was being critical of James A. Ford's work. Ford (e.g., 1962) did not use the terms we employ here, but early in his career (e.g., Ford, 1938b:11) he displayed a sophisticated awareness of the necessity of change being gradual when he noted that the act of sorting or ordering through seriation includes "overlapping." By this term he meant that type (class) A may fall within periods 1, 2, and 3, type B within periods 2, 3, 4, and 5, and type C within periods 4, 5, and 6. The various overlapping temporal occurrences of the types is what connects the sets of materials being seriated, ensures historical continuity and heritable continuity between collections, and allows an ordering to be derived. As Clement Meighan (1959:203) put it, "overlapping of similar traits [read *classes*] in different finds" is critical to the seriation method. If overlap does not occur, then one or more of the conditions or requirements of seriation will not have been met, and the seriation model will not be approximated.

Nels Nelson knew the value of overlapping traits. He had found superposed remains before the critical 1914 field season in the Southwest, but he noted that in such cases "there is often no appreciable [chronological] differentiation of remains" (Nelson, 1916:163). When he found evidence of chronological differentiation, it was between types at the ends of a continuum of several pottery types, and thus he lamented that such instances were "merely clean-cut superpositions showing nothing but time relations" (Nelson, 1916:163). However, when two types in the continuum were found stratigraphically mixed together, "one gradually replacing the other [, this] was the evidence wanted, because it accounted for the otherwise unknown time that separated the merely superposed occurrences of types and from the point of view of the merely physical relationships of contiguity, connected them" (Nelson, 1916:163).

This statement is important because it reveals that Nelson was thinking about culture change in terms that simply did not mesh with the thinking of many of his colleagues. In short, while they were thinking in terms of culture traits, the meaning of which was derived from ethnologically informed common sense, Nelson was thinking in terms of analytical units used to measure time. As a result, he replaced, at least in his own mind, the then-prevalent notion that culture change could be modeled as a flight of stairs, each step representing a static evolutionary stage and each riser representing a rather abrupt transformation from one stage to the next, with a model that viewed culture change as a gradually ascending ramp (e.g., Nelson, 1919a–c, 1932), albeit a ramp that moved through progressively more advanced stages. Plotting frequencies of pottery styles against time, which was rendered as geologically vertical space, would illustrate the gradual cultural evolution Nelson sought and eventually allow one to document the relative ages of the cultural stages. This was not only a revolution in analytical method, it was a revolution in metaphysic.

The Seriation Model

Historical types will, by definition, occur only during one portion of the temporal continuum and thus have what we refer to as a "continuous distribution." They will also display, during that period of occurrence, a unimodal frequency distribution relative to the abundances of other types. That is, the relative, or proportional, abundances of each historical type will initially be rare, eventually rise to a single peak abundance, not necessarily at the midpoint of its temporal duration, and finally decrease in abundance until it no longer occurs (Fig. 2.2). As George Brainerd (1951a:304) put it,

> if a series of collections comes from a culture changing through time [read *a tradition*], their placement on the time axis is a function of their similarity; collections with closest similarity in *qualitative or quantitative* listing of types lie next to each other in the time sequence. This ... allows a "seriation" or ordering of the collections to be formed which, if time be the only factor involved, must truly represent the temporal placing of the collections, although determination of the direction early to late must be obtained by other means. (emphasis added)

By "qualitative" we suspect Brainerd meant the presence–absence of types in collections, though this is not clear in his discussion; by "quantitative" we suspect he meant the relative frequencies of the types in collections. The former is referred to as occurrence seriation and the latter as frequency seriation. Note as well that Brainerd is correct when he says that "the direction early to late must be obtained by other means." Similarly, seriation, as we have said before, assumes nothing about the direction of change. Cowgill (1972:382) phrased it this way: "[F]or any specific sequence of entities, another sequence that is its exact reverse is an equally good seriation."

Both occurrence and frequency seriation assume transmission and heritability, and they do so at two levels (Rouse, 1939). First, each artifact identified as a member of a particular class is assumed to be related phyletically to every other specimen within that class, given their properties in common and typically (though not necessarily) their spatiotemporal propinquity (e.g., Phillips *et al.*, 1951; Rouse, 1955). Thus Brainerd (1951a:304) observed that a historical type "must be of sufficient complexity ... that the presence of an artifact belonging to [a historical type] suggests that its maker lived in the same cultural milieu as that of makers of all other artifacts classified into the same sorting group [read *historical type*]." We refer to this as the *type–species* sense of hereditary continuity (O'Brien and Lyman, n.d.). Second, the multiple classes that are seriated, whether by their occurrence or by their frequency, are assumed to be related phylogenetically given the requirement of seriation that all seriated collections derive from a single cultural tradition (Brainerd's "cultural milieu"). A "tradition" is, by definition, "a (primarily) temporal continuity represented by persistent configurations in single technologies or other systems of related forms" (Willey and Phillips, 1958:37). As such, a tradition reflects transmission, persistence, and heritable continuity (Phillips and Willey, 1953; Willey, 1945; Willey and Phillips, 1958). Because traditions can be conceived of and constructed at the scale of an attribute of a discrete object, a type of discrete object, or particular combinations of multiple types of discrete objects (e.g., Neff, 1992), we refer to this as the *tradition–lineage* sense of heritable continuity to signify the potential for a diversity of units, of whatever scale, within a tradition or lineage (O'Brien and Lyman, n.d.). Both occurrence seriations and frequency seriations are thought to indicate heritable continuity in both the type–species and tradition–lineage sense.

Requirements and Conditions of Seriation

If archaeological seriation is taken to be a method of comparing phenomena so that they may be ordered in such a manner as to reflect the passage of time, then the phenomena to be compared and ordered must be identical except for their position in time. The last means in part that the phenomena must be measured with theoretical units termed historical types or classes. It also means that the phenomena must meet certain conditions or requirements if a seriation is to produce a chronological order. Although the roots of the requirements are deep (e.g., Ford, 1936b, 1938a), one of the earliest and most detailed statements on them is found in Phillips *et al.* (1951:219–236). The requirements were variously expanded, amended, and clarified in later years (e.g., Cowgill, 1972; Ford, 1962; Rouse, 1967; Rowe, 1961). Robert Dunnell (1970) summarized them concisely, and we follow his discussion rather closely here.

First, assemblages of artifacts to be seriated must be of similar duration. Meeting this requirement insures that the positions of particular assemblages in an ordering are the result of their age and not their duration. What duration should be represented by the assemblages? As Ford (1962:41) noted, "Each collection must represent a short period of time—the shorter the better. A sampling of the [artifact] population representing an instant in time would be ideal but, of course, is never achieved." In other words, the shorter the duration, the finer the temporal resolution in the final seriation. Although it often is difficult to determine if, in fact, each set of material spans a similar duration of time, Irving Rouse (1967:162) indicates that "one can design [units] in such a manner that all will represent roughly equivalent periods of time." Ford's (e.g., 1935c:6, 1962:45) solution was to collect sufficient artifacts to ensure against variation in sample size per assemblage influencing the final order. Dunnell (1970:312) suggests that without absolute chronological control, which of course "obviates the need for a seriation," the best procedure is to attempt a seriation and determine if there are any sets of material that are "at substantial variance with the model stipulated by seriation." We provide an example of such later.

The second requirement is that all assemblages to be ordered must come from the same local area. This requirement is meant to insure that what is being measured is variation in time rather than difference in geographic space. It attends the fact that diffusion over geographic space can influence the results of a seriation (e.g., Phillips *et al.*, 1951:223). As Dave Davis (1981:57) notes, the use of seriation to measure time "relies upon general homogeneity in patterns of [artifact] change within a geographic area" and changes in "the nature and rate of inter-community contact" would obscure the patterns of artifact distribution expected by the seriation model. Thus, Rouse (1967:178) defines a "local area" as a chunk of geographic space "within which it is reasonable to suppose that there has been little, if any, geographic variation in culture." Thus, exactly what a "local area" is varies across space and through time (Dunnell, 1981). Only recently have techniques been developed that allow an archaeologist to determine analytically what a "local area" is in any given position along the spatiotemporal continuum (Lipo *et al.*, 1997). We describe these techniques later.

In our view, meeting the second requirement increases the probability of but certainly does not insure meeting the third requirement [following, particularly, Rouse's (1955) reasoning], which is that the assemblages to be ordered in a seriation all belong to the same cultural tradition. Given the definition of a cultural tradition (see above), if one meets this third requirement, then heritable continuity is assured and phylogenetic affinities between the seriated assemblages are guaranteed. As Dunnell (1970:311) notes, the third requirement means that the seriated assemblages "must be 'genetically' related" (see also Ford, 1938b). As Dunnell (1970:311) also notes, using theoretical units—classes—such as are demanded by both occurrence and frequency seriation, satisfies this requirement. That is, the use of classes of artifacts insures heritable continuity at the type–species level and, Dunnell contends, at the tradition–lineage level as well. With respect to the latter, the phylogenetic implications of the hierarchical structure of the Linnaean taxonomy in biology are transferable to a similar hierarchical alignment of artifacts. Thus, "pottery" might be aligned with a biological family, "types" of pottery with biological genera, and "varieties" of pottery with biological species, or the like. Classes of pottery can be seriated, as they comprise a pottery tradition, or monophyletic group, and projectile points comprise a different tradition, or monophyletic group (O'Brien and Lyman, n.d.). The two "families" of artifacts evolve independently of one another, and each therefore can serve as a test of the ordering produced by the other (Dunnell, 1970).

The conditions of occurrence and frequency seriation underpin the seriation model. Only by meeting the conditions will it be reasonable to infer that a seriation represents a chronology (Dunnell, 1970). One has to know then, if the conditions have been met. Given that the seriation model holds that "differences between units mainly reflect differences in time, and that the seriation sequence is a good approximation to the time sequence of the units" (Cowgill, 1972:383), failure to meet the conditions of the model will result in an ordering that may not reflect the passage of time. Before we indicate how one can determine if those conditions have been met, we need to present the occurrence and frequency seriation techniques in more detail. Although frequency seriation was invented first, we begin with occurrence seriation because it is the simpler of the two and its basic analytical principle is also part of frequency seriation.

OCCURRENCE SERIATION

Occurrence seriation was suggested as an alternative to frequency seriation in the late 1950s and early 1960s; the first extended discussion of it of which we are aware is a paper by Paul Dempsey and Martin Baumhoff (1963), though John Rowe (1959) mentioned it several years earlier. As William Lipe (1964:103) noted, the major difference between what Rowe and Dempsey and Baumhoff proposed and what previous researchers such as Ford (e.g., 1949, 1951) and Brainerd (1951a) had done was that "with the former, collections are compared in terms of presence or absence of types, whereas in the latter, the relative frequencies of types within the respective collections are employed in the comparisons." Rowe (1959: 321) argued that "It is preferable to avoid relying on frequencies [of types] for making chronological distinctions and depend instead on observations of presence and absence," because frequency data are subject to sampling problems. Dempsey and Baumhoff (1963:498) also believed that presence–absence data would be more sensitive to chronological issues because "types that occur with low frequency may be among the best time-indicators [and] the presence of single specimens of certain types may be crucial in establishing chronologies." They also found that by "weighting" a type's importance or role in a seriation on the basis of its relative abundance, though objective, resulted in "gross differences in the contribution of the various [types] to the final ordering" and "the chief effect of [including rare types] is to increase the amount of busywork" (Dempsey and Baumhoff, 1963:498).

To circumvent such problems, Dempsey and Baumhoff (1963:501) recorded each type "merely as being present or absent" in each of the collections they wanted to order. Rather than summarize the rather complex archaeological case they describe, we offer Table 4.2 as a simple example of the occurrence seriation procedure. The example consists of six assemblages, the artifacts of which have been classified as belonging to five historical types. The seriation model stipulates that "the distribution of any historical or temporal class is continuous through time" (Dunnell 1970:308). This is in short the principle on which the occurrence seriation procedure rests. The procedure is to sort the unordered rows—the collections—in Table 4.2 such that each type (each column) displays a continuous occurrence, or column of "+" marks. The order resulting from meeting the expectations of the seriation model is given in the ordered part of Table 4.2. Note

Table 4.2. An Example of an Occurrence Seriation Procedure

Assemblage	Historical type 1	2	3	4	5
Unordered					
A	+		+	+	+
B			+	+	+
C	+		+		+
D			+	+	
E	+	+			+
F			+	+	
Ordered					
E	+	+			+
C	+		+		+
A	+		+	+	+
B			+	+	+
D/F			+	+	

that it makes no difference if the ordering from top to bottom is "E, C, A, B, D/F" or "D/F, B, A, C, E," because the direction of time's arrow is unknown. That knowledge must come from other data independent of the seriation, such as knowing that type 1 and 2 occur late in time and types 3 and 4 occur early in time, based on associated radiocarbon dates or stratigraphic excavations.

We note that in Table 4.2 assemblages D and F are identical in terms of the types they contain. They cannot be sorted and must, in this example, be considered *contemporaneous*. This is a slippery concept in archaeological dating. Although in our (fictional) example assemblages D and F may have been formed and deposited at precisely the same instant in time, this is unlikely. By saying that the two assemblages are contemporaneous we mean only that they cannot be distinguished in terms of the units with which we have chosen to measure time in the case at hand (see also Patterson, 1963). Had units other than the presence–absence of types been used, perhaps the two assemblages could be separated.

Occurrence seriation is an inexpensive and relatively easy dating technique, yet it has not been used very often. One scholar who has used it with great effect is Michael Graves. He produced chronologies of pottery designs in the American Southwest (Graves, 1982, 1984) and chronologies of architectural features in Oceania (Graves and Cachola-Abad, 1996), which have supplemented chronological information produced with dendrochronological data and radiocarbon dates. His analyses allow him to address issues of contemporaneity of design styles and perhaps most important, to conclude that change in design styles was "incremental" and that rapid design replacement reflects a disruption of "the orderly transmission of information across generations" (Graves, 1984:17), or, in our terms, a disruption of the mechanism of hereditary continuity. Other conclusions he derives from his occurrence seriations concern social structure and population interaction, illustrating the utility of such seriations for archaeological inquiry above and beyond the mere construction of a relative chronology. His papers warrant close study.

FREQUENCY SERIATION

In addition to the stipulation that "the distribution of any historical or temporal class is continuous through time" (Dunnell, 1970:308), the frequency seriation model specifies an additional principle. This additional principle is a simple one, elegantly stated by Phillips, Ford, and Griffin (1951:220) with respect to pottery types constructed for the central Mississippi River valley:

> If our pottery types are successful measuring units for a continuous stream of changing cultural ideas, it follows that when the relative popularity of these types is graphed through time, a more or less long, single-peak curve will usually result. Put in another way, a type will first appear in very small percentages, will gradually increase to its maximum popularity, and then, as it is replaced by its succeeding type, will gradually decrease and disappear.

The curves that Phillips, Ford, and Griffin describe are the ones shown in Fig. 2.2. They noted that if "a complex of cultural materials representing a space–time continuum of culture history is classified in a consistent manner, the popularity curves of the various constituent types will form a pattern. Each portion of this pattern will be peculiar to a particular time and area" (Phillips *et al.*, 1951:221).

In their discussion, Phillips, Ford, and Griffin (1951) likened the seriation of pottery sherds to reconstructing the history of transportation in Ohio, which we reproduce in Fig. 4.1. Modes of transportation are arranged vertically, the direction in which time is running. Horizontal bars represent percentages of each mode of transportation at years divisible by ten. Summing percentage bars across any row always yields 100%. Thus, in 1940, automobile transportation accounted for 51% of all transportation in Ohio, rising from 42% in 1930. Similarly, horse-drawn vehicles accounted for only 2% of transportation in 1940, down from 6% in 1930. Note that the relative frequencies of all transportation modes differs by 10-year periods; as Phillips, Ford, and Griffin pointed out, the quantitative picture of any particular block of culture history differs from all others. Thus one could determine the correct chronological position of "sample X" from Ohio based on the frequencies of each type of transportation, as shown in Fig. 4.1. Importantly, the pattern for Ohio at any particular point in time will differ from patterns obtained elsewhere.

The distributional diagram for this imaginary example works the same when

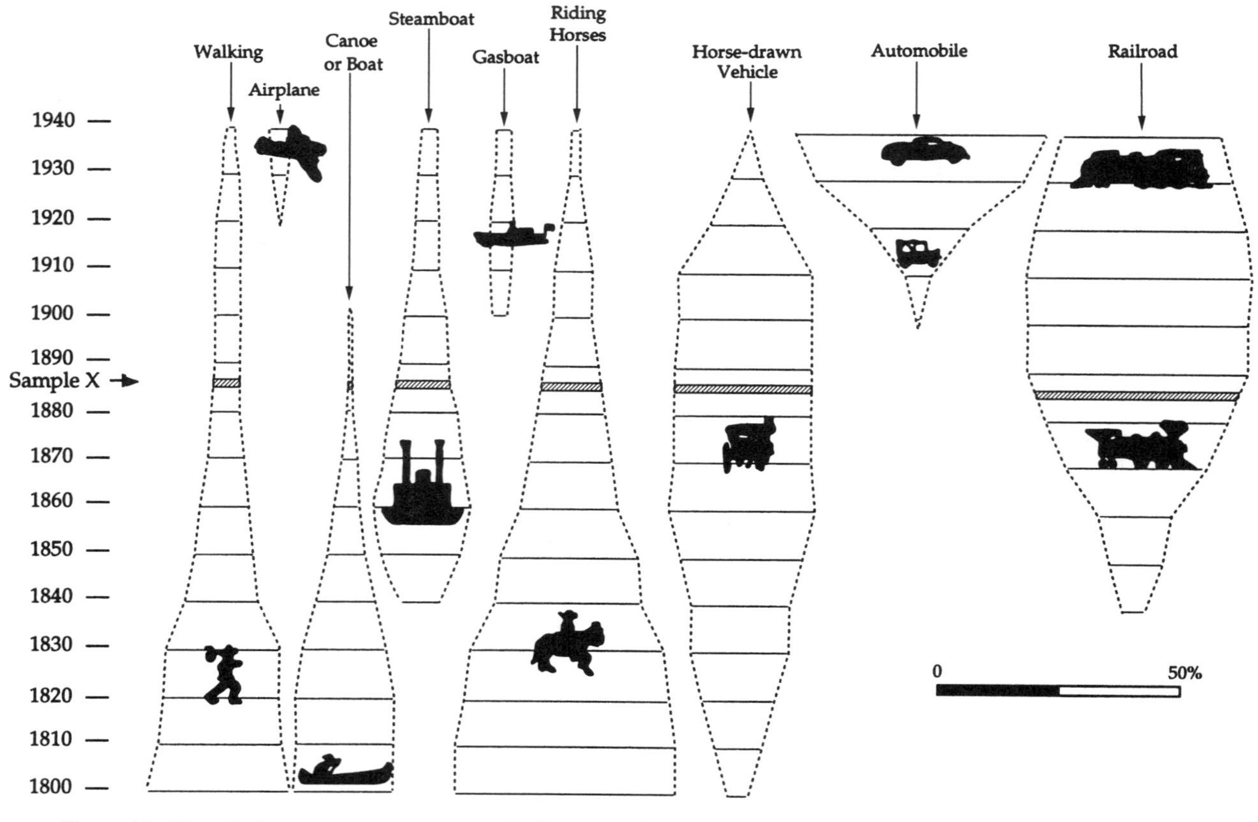

Figure 4.1. Theoretical percentage occurrence graph of transportation types in Ohio from 1800 to 1940 (after Phillips *et al.*, 1951).

Table 4.3. An Example of a Frequency Seriation Procedure

Assemblage	Historical type 1	2	3	4	5
Unordered					
A	10		30	10	50
B			50	30	20
C	20		15		65
D			40	60	
E	30	25			45
F			20	80	
Ordered					
E	30	25			45
C	20		15		65
A	10		30	10	50
B			50	30	20
D			40	60	
F			20	80	

archaeological specimens are used. As the example in Table 4.3 shows, historical types are aligned vertically. Each row represents a single collection of specimens, each from a distinct spatial position on the landscape (not to preclude multiple collections from different areas of a single site), and as in the imaginary example of transportation modes, the percentages of the various types in each collection sum to 100%. What Phillips, Ford, and Griffin showed was a completed seriation that was constructed to illustrate the frequency seriation model. Our example in Table 4.3 begins with the collections in no particular order. The seriation procedure would be first to rearrange the rows such that each type meets the historical continuity principle. This has already been done in Table 4.2, as the Table 4.3 data comprise the same (fictional) collections as the former but with frequency data rather than mere presence–absence data. The second step in the seriation procedure is to sort the collections such that each column of frequencies defines a unimodal frequency distribution, such as is shown in the ordered part of Table 4.3. Note that as with occurrence seriation, it makes no difference if the ordering from top to bottom is "E, C, A, B, D, F" or "F, D, B, A, C, E"; the direction of time's arrow is still unknown. Again, knowledge of the direction taken by time's arrow must come from other data independent of the seriation.

Note as well that in the (fictional) example in Table 4.3, assemblages F and D are not contemporaneous, as they are in Table 4.2. The relative frequency data allow these two assemblages to be separated and ordered along with the other included assemblages. Although such increased resolution in temporal ordering

may not always be found with frequency data relative to presence–absence data derived from the same collections, we suspect it often will be because of the greater information content in the former compared to the latter. And, as Steven LeBlanc (1975:23) noted, "chronologies based on the presence or absence of types or attributes will never be as accurate as those based on their relative frequencies. This is because technological and stylistic changes are rarely instantaneous in inception or momentary in duration." What LeBlanc meant was that change will occur as shifts in the particular combinations of variants or the frequencies of variants, whether attributes of discrete objects or types of discrete object, rather than variant A occurring only prior to one point in time and variant B occurring only after that particular point in time. This is merely another way of saying that change is gradual and involves overlap.

Our example in Table 4.3 is a simple one; in reality, archaeologists often have more than five historical types and more collections than the six in the example. Suppose we had pottery collections from 50 sites and we wanted to perform a seriation. How would we go about doing it? There are numerous computer programs that seriate collections automatically and produce best fits given the parameters that the analyst inputs. We do not delve into these [see references in Johnson (1972) and Marquardt (1978) and the discussion in Lipo *et al.* (1997)], because what we believe is more important in the present context is to understand the principles of seriation. As we document elsewhere (Lyman *et al.*, 1998), archaeologists spent considerable time during the first half of the twentieth century trying to determine the best way to present both the data and the results of a seriation. The "best way" was found to comprise a centered bar graph that showed not only the relative frequencies of the types but how those frequencies changed through time. Informally known as the "Ford technique," this manner of presenting data and results simultaneously

> presents far more information than do the comparable [and more modern] statistical techniques. Not only is the degree of similarity between two units indicated, but the actual form and source of similarity is shown as well. A [statistically based] seriation can always be constructed from the information contained in a Ford seriation, but graphic seriations cannot be constructed from a matrix of similarity coefficients. (Dunnell, 1970:306)

Thus, before computing power was available, archaeologists did frequency /seriations by hand. Ford (1962) suggested using long strips of paper containing bars of length proportionate to the percentage of a particular pottery type in a particular collection. Each strip shown in Fig. 4.2 is a separate collection with bars showing the percentage of each type of pottery contained in that collection. Eleven pottery types are shown on the graph, though no collection contains sherds of all 11 types. Once each collection is graphed in terms of type percentages, the strips are moved up and down until a best fit is found, meaning that there are as few violations of the continuity and the popularity principles as possible. That is, the

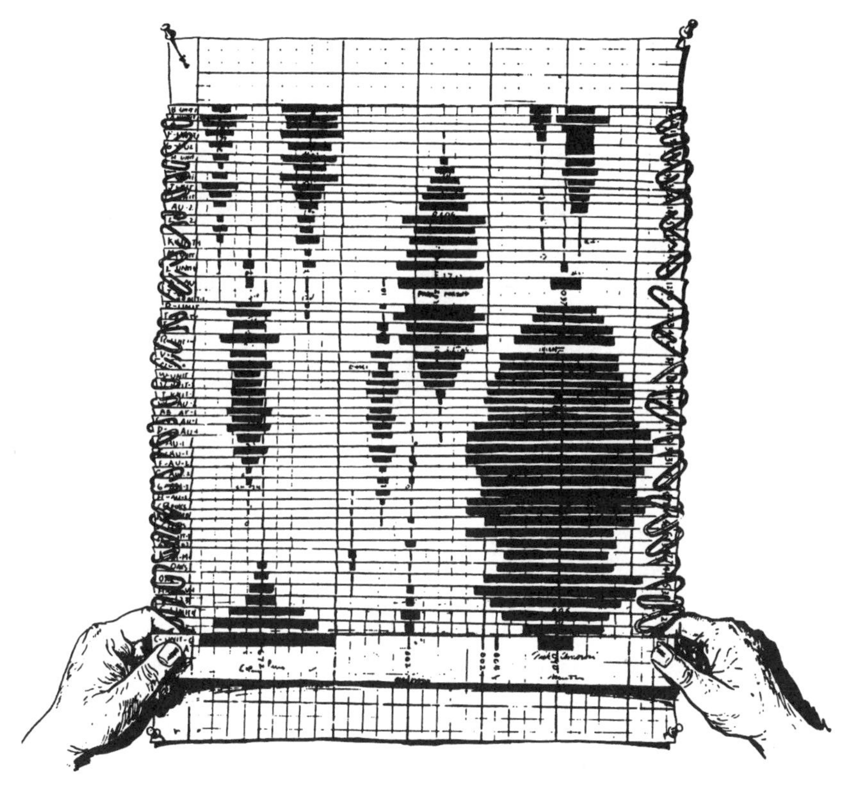

Figure 4.2. James A. Ford's thumbs-and-paper-clips method of seriating collections. Each strip of paper represents a surface collection or excavation level; on each strip bars have been drawn to indicate the percentage of each pottery type. The strips are then moved around until the best fit is attained (from Ford, 1962).

resulting type frequency curves, what Ford (1952:344) referred to as " 'battleship' frequency curves," are as close as possible to those shown in Fig. 2.2. There are several reasons why such curves may not be found, and we turn to them next.

MEETING THE CONDITIONS OF THE SERIATION MODEL

Slight deviations from the ideal battleship-shaped curves, a few of which are noticeable in Fig. 4.2, can perhaps be explained by sampling error. What does that mean? Archaeologists have long realized that not all collections of artifacts meet the requirements of seriation. With respect to occurrence seriation, Cowgill (1968:518) indicated that it made sense to perform such a seriation only "when we feel sure that the absence [of a type] in the collection means that the [type] was *really* absent" and not the result of sampling error. Ford (1962:41), writing about frequency seriation, indicated that "Each collection must be unselected and large enough to give reliable percentages. A total of 100 or more sherds is desirable, but occasionally collections as small as 50 can be used." Ford's imposition of the 50- to 100-sherd sample was based on his experience in the Mississippi Valley, which had shown that samples of that size often yielded adequate results, but the size of an adequate sample size will probably vary from place to place. He was right on the mark with his comments about "unselected" samples being desirable, just as Kroeber had been over four decades earlier, since anything less than a near-random sample would introduce bias into the analysis. Archaeologists now have many ways of assessing sample adequacy, and we need not elaborate on them here (see Leonard and Jones, 1989, and references therein). It suffices to note that when performing either an occurrence or a frequency seriation the analyst should check for sample size effects; to date, such efforts have been rare in archaeological applications of the seriation method (see Lipo *et al.*, 1997, for an exception). What of the other, more formal requirements of the seriation method?

Cowgill (1968:517) noted that seriation can be conceived of as having two "central tasks": (1) determining the correct chronological sequence of a set of archaeological units, and (2) ordering units based on their similarity. Importantly, Cowgill (1968) emphasized that the best ordering in terms of unit similarity may or may not also be a chronological ordering. Thus, in order for the accomplishment of the second task to result in the simultaneous accomplishment of the first task, an ordering produced by occurrence or frequency seriation can be inferred to be chronological only if the conditions of the seriation model have been met. How then do we determine if those conditions have been met? The first condition is that all collections must be of comparable duration. Dunnell (1970:312) suggested that for a frequency seriation, if a unimodal frequency distribution for all included types cannot be obtained, then perhaps the collections vary in duration. Further, collections of overly long duration, termed *mixed* assemblages by Phillips, Ford, and Griffin (1951) and others, may contain types from rather different vertical positions in the seriation. An example of such a collection is given in Fig. 4.3. In this example, both the principle of continuous occurrence and the principle of unimodal frequency distribution are violated, the former by type 4 and the latter by type 2. Such deviations from the seriation model cannot be tolerated, particularly when they are found in the middle of the duration of a type: "Variation from the [seriation] model can be tolerated only where it is predictable, namely at the beginnings and ends of distributions and in sparsely populated classes" (Dunnell, 1970:313).

Another requirement of the seriation model is that the seriated collections come from the same local area. This condition is the most difficult one to meet (Dunnell, 1981). As James Deetz and Edwin Dethlefsen (1965) showed, artifact

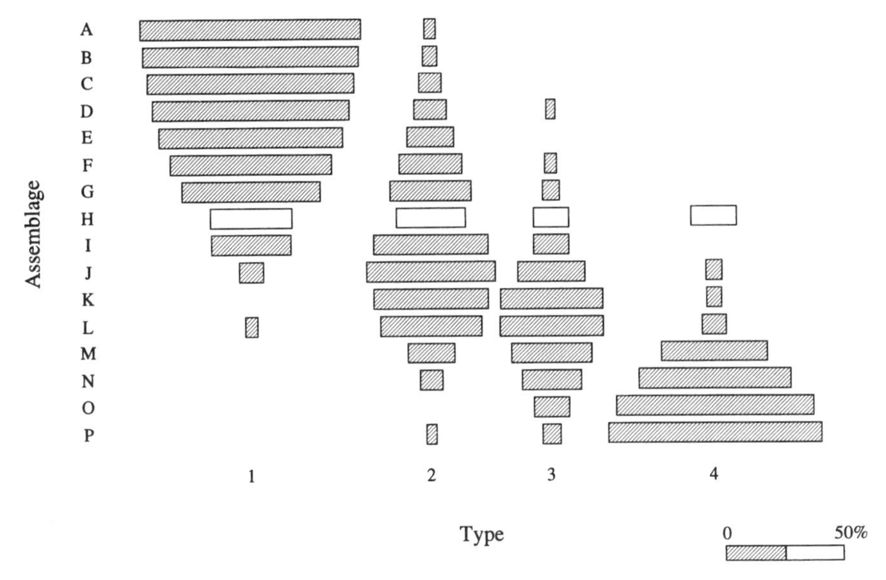

Figure 4.3. A seriation of 16 assemblages and four artifact types showing the effect on the expected frequency pattern of an assemblage (H, unshaded) of different duration (longer) than all other assemblages in the ordering (after Dunnell, 1970).

form can vary more or less continuously across space, so the problem becomes one of where to draw a boundary line around a "local area" such that formal variation within the bounded area is minimal and between such areas is maximal. Another way of saying this is that both "within-spatial-unit homogeneity" of form and "between-spatial-unit heterogeneity" of form are to be maximized. There are at present two ways to attempt to meet this requirement. One might construct historical classes that vary greatly along the temporal dimension and minimally along the spatial dimension; that is, the classes should consist of attributes that "show little variation in space and much variation in time" (Dunnell, 1970:315). Thus Cowgill (1972:384) noted that "it is only necessary that [nontemporal] sources of variation, within the particular set of units being studied, make for differences that are small relative to differences reflecting the smallest time intervals one hopes to reliably distinguish."

The other way to meet the "same local area" requirement is empirical and rests on the notion of heritable continuity in the tradition–lineage sense. That is, it is explicitly founded in a Darwinian notion of heritability realized by transmission, in this case, of ideas. Under such a conceptual umbrella, one can argue that a "local area" comprises a "community or localized group of communities [that] produces a distinctive style of pottery that is distinguished easily from the products of other centers" of communication and pottery production (Neff, 1992:151). Following this notion, and using the principles of frequency seriation and the transmission model developed by Fraser Neiman (1995), Carl Lipo and his colleagues (Lipo *et al.*, 1997) simulated trait transmission and mixture over time and across space. They found that perfect battleship-shaped curves could be generated, with sufficient control of time, using collections recovered from particular limited pieces of geographic space. Applying their findings to archaeological data from the Lower Mississippi Alluvial Valley, they identified instances of prehistoric community interaction and structure. That is, they identified a set of "local areas" on the basis of the empirical similarities of the collections. Of course, because cultural transmission pathways can alter direction over time, in other words, cultural transmission and thus evolution is reticulate, the analyst will need to determine if the "local areas" identified for one portion of the temporal continuum change their boundaries over time. Given our understanding of cultural transmission, we suspect they will. This in turn demands that the spatial dimension be monitored and controlled, so-called "local areas" identified, throughout a seriated set of collections.

The final requirement of the seriation model is that the seriated collections all belong to the same cultural tradition. The easiest way to recognize a failure to meet this condition is when, in a completed seriation, there is marked discontinuity or a lack of overlap among the collections. An example is given in Fig. 4.4. Such a result may be produced by the representation of more than one "local area" by the collections that are being seriated; this possibility can be checked quickly by determining if, in the case of Fig. 4.4, assemblages A, B, C, and D come from one area and assemblages E, F, and G from another, and the two areas do not overlap in space. If the two areas identified by the assemblages do overlap in space, then it is possible that the two sets of assemblages represent a distinct break in heritable

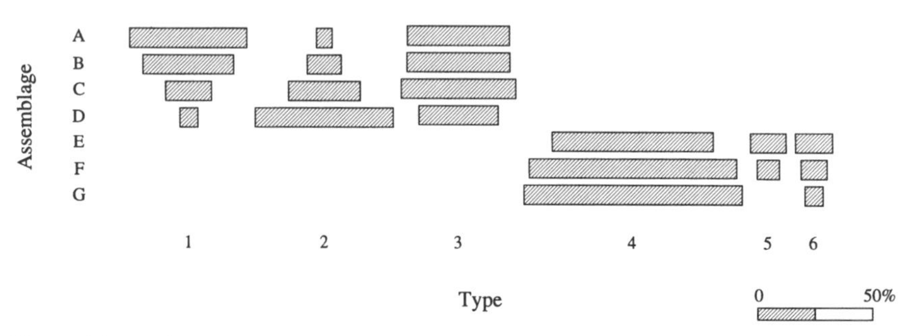

Figure 4.4. A seriation of seven assemblages and six artifact types showing the effect on the expected frequency pattern of including two traditions between which there is no heritable continuity or overlap. Assemblages E, F, and G could legitimately have been placed on top rather than the bottom (after Dunnell, 1970).

continuity, and thus two separate traditions. The example in Fig. 4.4 is a clear case of a violation of the principle of heritable continuity, of historical continuity, and of Cowgill's (1972:384) requirement that there never "be a break in the sequence so abrupt and catastrophic that units immediately following the break bear no (or only accidental) resemblance to units before the break." In other words, there is no overlap, historical or heritable, to connect the two sets of assemblages.

The preceding should not be confused with the patterns of frequency curves that emerge from a frequency seriation. As Dempsey and Baumhoff (1963:497) note, there are minimally four patterns that may appear in a frequency seriation graph; these are shown in Fig. 4.5. The patterns emerge from various aspects of the model of seriation in combination with aspects of archaeological sampling. First is the popularity principle, or the notion that every artifact type will display a unimodal frequency distribution over time. Second, different types will have different histories, meaning that they will (1) span different durations of time and (2) have different relative frequency histories. And third, seldom will the entire lineage or tradition be represented by the set of collections being seriated. The first two concern the seriation model; the last concerns archaeological sampling. In combination, these aspects mean that some types will appear at either end of the seriated order, but their complete historical duration will not be represented by the available sample of assemblages (patterns I and IV). Other types will span the

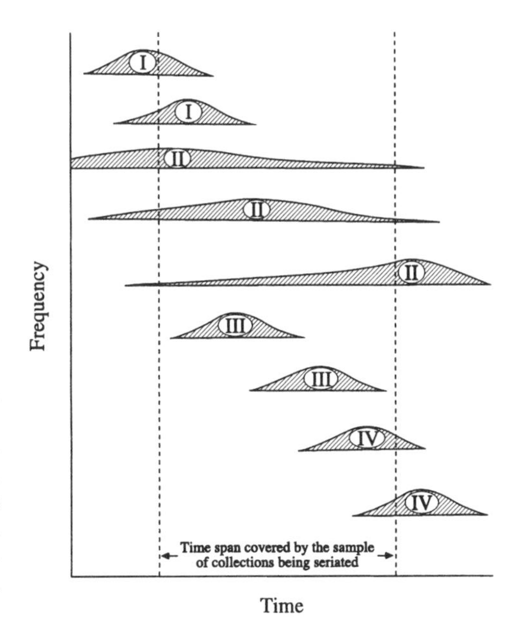

Figure 4.5. Kinds of frequency distribution patterns (I–IV) possible in a sample of sites falling within the time period defined by the dotted lines. A historical type may occur in all site-specific assemblages included in the seriation (pattern II), at only one end of the seriation order (patterns I and IV), or only in the middle of the order (pattern III) (after Dempsey and Baumhoff, 1963).

entire duration of the sample of collections, yet their complete history will not be represented by the available sample (pattern II). Finally, there perhaps will be some other types the entire history of which is represented (pattern III). Finding such cases is not indicative of any violation of the seriation model. Dempsey and Baumhoff (1963) were concerned that types displaying pattern III would introduce errors in the ordering and thus in the chronology, but they were wrong (Cowgill, 1968). What their discussion does do, however, is to introduce, if indirectly, the degree of temporal resolution provided by seriation.

TEMPORAL RESOLUTION AND RATES OF CHANGE

As we noted earlier, a good ordering of collections [Cowgill's task 2 (1968)] may not be a good chronology [Cowgill's task 1 (1968)]. One reason why the best ordering may not reflect the passage of time involves the fact that the units within the seriation may not be temporally sensitive (Cowgill, 1968:519). But, presuming that a good ordering can be shown to be a good chronological ordering, such as with, say, stratigraphic information, then the next question to ask concerns the degree of temporal resolution provided by the ordering. Some believe that the degree of temporal resolution obtained will depend on the rate of change within the lineage or tradition represented by the seriated materials. Clement Meighan (1977:628), for example, notes that "if there is rapid [cultural] change, any orderly seriation method will recognize short time periods; with data showing little observable change over long periods, no seriation method can identify short time periods."

Of course, the problem described in the preceding paragraph concerns sortability, meaning how easily collections can be distinguished from one another, which in turn rests on the variables used to describe them. In the case of frequency and occurrence seriation, the variables are the frequencies or presence–absence of types, respectively. This means the problem reduces to one of classification or the construction of types: Which variables or attributes should be used? Meighan (1977:629) thus correctly noted that "Sometimes the quality of the data [used in a seriation] can be improved merely by improving the definition of the ceramic types ... so that the types more closely reflect chronological change." Meighan (1959: 203) suggested that each type "should occur throughout the time span studied"; that is, each type should display Dempsey and Baumhoff's pattern II (Fig. 4.5). By now it should be clear that this is nonsense. LeBlanc (1975:24) suggested that "too few types do not distinguish short intervals of time," so one might think more types would be better. But as Cowgill (e.g., 1968) and others have noted, the more types (and assemblages) included in a seriation, the more difficult it becomes to produce a clean ordering, one that does not violate the continuity and unimodal frequency principles.

If types are historical and are defined such that they existed over some neither exceptionally short nor excessively long duration of time, then they should be seriable. That is, they should not be so limited in time that they occur in only one assemblage but in no others; were such the case, then no overlap between assemblages could be detected and continuity would be precluded. But the types also should not occur over so long a time span that in, say, an occurrence seriation every type is found in every assemblage. In such a case no sorting and thus no ordering of the assemblages can occur. Similarly, we suspect that if types occupy spans of time that are of exceptionally long durations, then it will be difficult to derive an ordering in which the types display perfect unimodal frequency distributions. LeBlanc (1975) suggested that in some cases frequency seriation of particular attributes, rather than attribute combinations typically termed types, may provide an ordering that more closely approximates the model of seriation, and thus represent finer temporal resolution. He referred to this as "microseriation." In the sense that temporal resolution may in fact be finer, we agree with his term. But, "a combination of a larger number of [attributes] is likely to cover a shorter span of time than a combination of a smaller number" (Rowe, 1959:322). That is, each attribute, as a class, will have a particular duration, and there is no reason to suspect a priori that each attribute will have a duration that overlaps perfectly with any other attribute. Thus, a combination of attributes that we choose to call a type will have a shorter duration than any one of its constituent attributes, but only if the combination of attributes is established independently of the duration of any given attribute. Such would be easy to determine empirically in any given situation: merely seriate collections based on types, then based on attributes, and compare the two. Individual attributes should span longer durations, occur in more assemblages, than individual types.

Along lines similar to the preceding, Frank Hole and Mary Shaw (1967:96) indicated that "a comparison of different data sets from the same series of sites gives dramatic evidence of relatively different rates of change among data sets. These data, graphically expressed, could go a long way toward making the nature of change more understandable." That is, projectile points, say, will have a history of change different from the history of change in pottery, even though both belong within the same cultural lineage. This is so because, as Hole and Shaw (1967:86) indicated, types that produce good seriations "are likely to have a relatively neutral adaptive value in a culture. Changes in them will derive more from fashion or from random drift than from necessity" (see also Dunnell, 1978; Lipo *et al.*, 1997; Neiman, 1995). In other words, change in projectile points may be slow (or fast) relative to change in ceramics; the two will change independently of one another. Such a (fictional) example is shown in Fig. 4.6. This returns us to Meighan's (1977) comment regarding rates of change that began this section. More rapid change in the ceramic tradition shown in Fig. 4.6 (top) is indicated by the turnover of types from the bottom to the top of the graph as well as by the lack of stability of

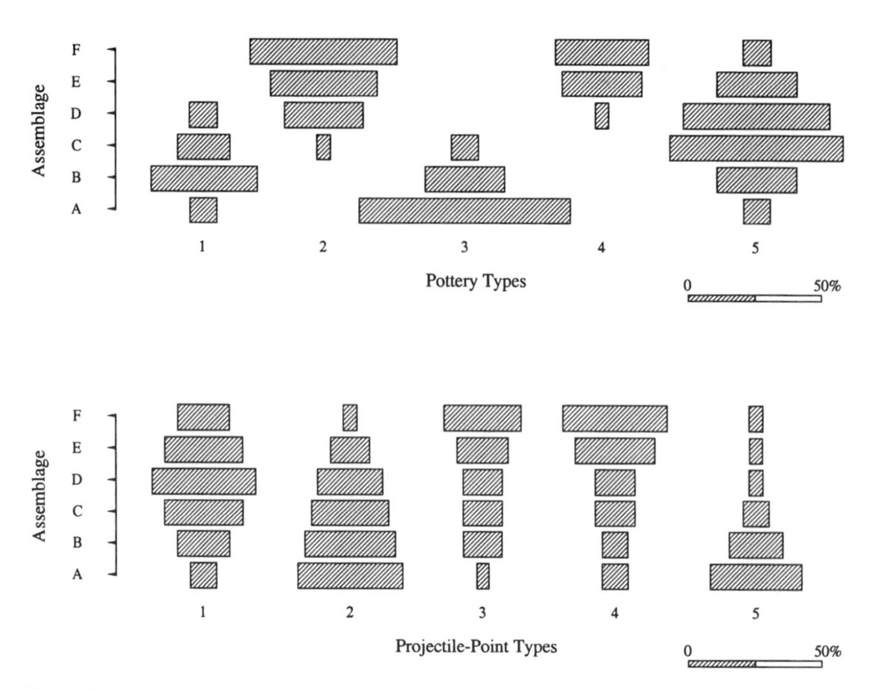

Figure 4.6. Comparison of a seriation order of six assemblages based on pottery (top) with a seriation order of those assemblages based on projectile points (bottom). Pottery is changing rapidly relative to the gradual changes in projectile points.

frequencies of some types across the graphed collections. In contrast, the projectile point tradition shown in Fig. 4.6 (bottom) includes all types throughout the graph, and the relative abundances of projectile point types 3, 4, and 5 do not change across two or three assemblages. Such information could prove valuable, as Hole and Shaw noted, for revealing the nature of change.

Absolute Seriation

In a literal sense, the degree of temporal resolution afforded by any occurrence or frequency seriation is difficult to evaluate without some interval scale measure of time against which to align the ordering. If such a scale is available, then seriation may not be required to produce a temporal order. Seriation is a powerful ordinal scale method for ordering phenomena chronologically, because, as Dunnell (1981:67) points out, it can be applied in a wider variety of circumstances than can most other chronological methods. As we have seen, there are numerous seriation techniques one can use to order phenomena, none of which depends in the slightest on stratigraphy. Variation in any of several properties of specimens (form and decoration, for example) provides an excellent basis for marking the passage of time (phyletic and evolutionary seriation) as do changing percentages of specimens placed in types (frequency seriation) or even the presence or absence of particular attributes or types (occurrence seriation). An ordering produced by seriation is just that, an ordering. Archaeologists hope that the ordering also represents a sequence, that is, a chronology, but this must be tested with other evidence independent of the seriation. But even a correct temporal sequence tells us nothing about the amount of time that elapsed during production of that sequence. We need an independent source of such information. In biology we have access to molecular clocks, which are based on the premise that genetic mutation rates are fairly constant. Thus if organisms in two species are 96% identical in terms of their DNA, and DNA mutates at the rate of 0.5% every million years, then the species have been separated for roughly 8 million years. Archaeology has a host of independent dating methods that can be used for similar kinds of control.

There is, however, no obvious reason to assume that rates of change outside the molecular realm are constant. In fact, there are very good reasons to believe they are not. Slow, gradual change can be modeled as shown in Fig. 4.7, where after a while enough change in characters and character states has taken place that a new entity, for example, a species, appears. But what if change is not spread out evenly over time? What if there actually are long periods of inactivity (stasis) punctuated

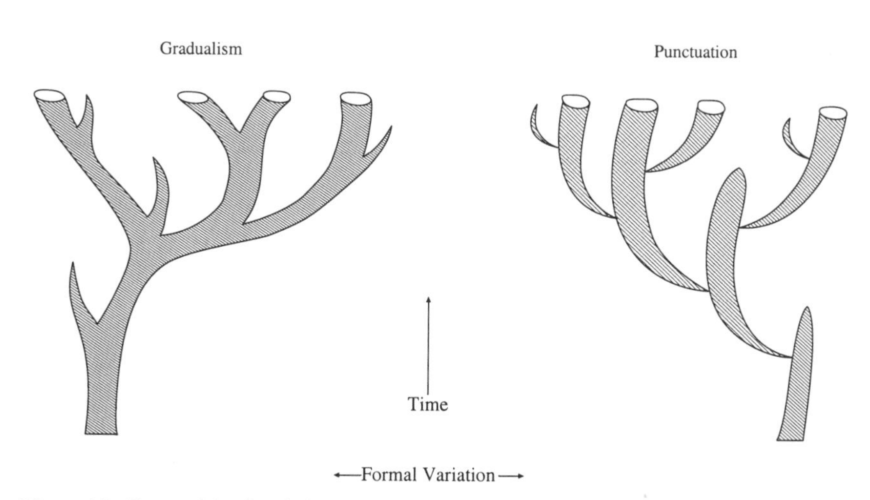

Figure 4.7. Two models of evolutionary change showing one parent species giving rise to four extant species (top of the chart equals present day). Gradualism (left) views evolutionary change proceeding steadily through time, whereas punctuationism (right) views most of the change as being concentrated in relatively short bursts (after Lewin, 1982).

by sudden and short bursts of rapid activity? The end result, production of a new entity (Fig. 4.7), is the same as that produced as a result of a slow buildup of new character states. The two processes, however, are markedly different. In biology, the rapid burst of activity that leads to speciation is called *punctuated equilibrium*. This concept has yet to be applied systematically to the archaeological record, but there is nothing on the face of it that indicates an inappropriateness to the study of human phenotypic features and the timing of the appearance of those features.

There have been a number of attempts to convert the ordering produced by a seriation into an absolute dating tool. In short, the procedure involves correlating the ordinal scale ordering of materials with an interval scale measure of time, the latter often being provided by radiocarbon dates associated with materials typologically identical to those in the seriation. Reasons given for doing so include the greater chronological sensitivity of interval scale chronologies. For example, Robert Drennan (1976:292) noted that one problem with seriations is that "they simply order [artifact] lots without providing any indication of the temporal spacing of units within the ordering." Similarly, Dwight Read (1979:91–93) stated that frequency seriation places assemblages into "a one-dimensional, ordinal sequence [that] is deceptive" because at least some of the assemblages might appear contemporaneous given such a scale of resolution, and that "changes in the proportion of [artifact] types in sites is discontinuous" rather than continuous. He therefore urged the construction of "finer time frameworks for interpreting archaeological data." That is, he wanted a way to objectively identify the discontinuities of time and to sort out which assemblages were actually contemporaneous from those that only appeared to be so. This demanded in part an absolute or interval scale chronology.

David Braun (1985:509) sought a calibration method to "produce a continuous, accurate estimate of absolute age along the seriation scale" and referred to it as *absolute seriation*. He plotted the thickness of pottery sherds against a series of associated radiocarbon dates to derive a chronocline, or character gradient, of pottery thickness on which he could superimpose interval-scale dates (Braun, 1987). Such a procedure allowed detection of the rate of change. An earlier but similar study involved measuring the diameter of the hole in clay pipe stems. J. C. Harrington (1954) measured, to the nearest 1/64th inch, the diameters of the stem holes in a sample of pipes made between AD 1620 and AD 1800. He presented the results in graphic form, redrafted in Fig. 4.8 as a frequency seriation. This is a rather clean frequency seriation. Lewis Binford (1962) found it difficult to use Harrington's chart, so to remove the difficulty he converted Harrington's data into an absolute scale by calculating a simple best-fit regression line between the average hole diameter per temporal period and the midpoint of each period. The regression formula describing the line is

$$Y = 1931.85 - 38.26X$$

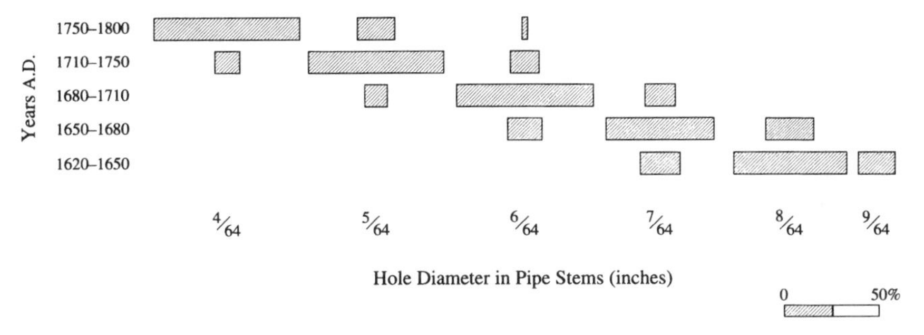

Figure 4.8. A frequency seriation of classes of hole diameter in pipe stems based on data collected by Harrington (1954).

where *Y* is the estimated age of the unknown pipe specimen, and *X* is the observed diameter of the stem hole in the unknown specimen.

There subsequently was some debate as to the validity of Binford's dating technique (e.g., Binford, 1972; Hanson, 1971, 1972, and references therein). For one thing, it presumes that the rate of change in the character being measured, in this case, the diameter of the stem hole, was constant over the time span represented; Hanson (1971), for example, showed the statistical relation was better described as a curved rather than a straight line. Regardless of whether the relation between calendric age and stem hole diameter is best described by a straight or curved line, two important points emerge. First, any such statistically derived line, of whatever shape, is an empirical generalization, the shape of which will depend on the specimens used in the derivation. Second, the choice of line to fit to a data set demands an assumption about the rate of change: was it stable, did that rate change, and if so how. Any such assumption becomes progressively less reasonable as the time span incorporated in the statistical calculation increases in duration (e.g., Braun, 1985, 1987).

Yet, some archaeologists argue that such regression analysis is desirable. Stephen Plog and Jeff Hantman (1990:444) indicate that if

> accurate [regression] equations can be developed, the use of regression analysis has advantages over more commonly used seriation techniques. Not only can sites without independent dates be placed in chronological order but estimates of the actual occupation date of each site also can be calculated along with a confidence interval for that estimate.

In making this statement, Plog and Hantman (1990:439) seem to miss their own most important observation, which is that archaeologists "must constantly confront the epistemological dilemma created by the need to simultaneously impose and interpret temporal patterns." Rather than heed this warning, they derive a

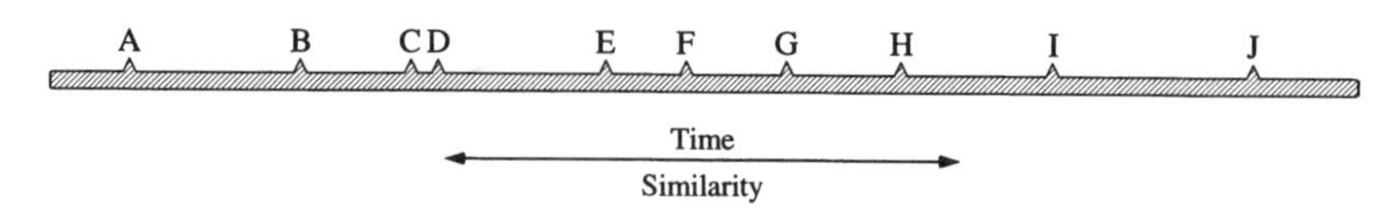

Figure 4.9. A sequence of ten collections (A–J) arranged according to their similarity. Collections more similar to one another are arranged closer together and are closer in time. However, whereas similarity is measured on an interval scale, time is measured on an ordinal scale, and thus the arrangement is useless for measuring absolute temporal differences between the ten collections (after Dempsey and Baumhoff, 1963).

regression equation and then use it to date sites that are otherwise of unknown age (Plog and Hantman, 1986, 1990). That is, they derive an empirical generalization regarding a temporal pattern, then impose that pattern on unknowns, and finally interpret the result.

A case in point of attempting to render interval scale measures of time from seriations is found in Dempsey and Baumhoff's (1963) measures of similarity based on the types shared by the seriated assemblages. Noting that the basic seriation model holds that "the pattern of artifacts found at a given site will be most similar to the pattern of sites near it on the time dimension and most different from the patterns at greatest remove in time," Dempsey and Baumhoff (1963:502) produce a figure in which they show a "hypothetical sequence of ten sites." We show a version of that sequence in Fig. 4.9. Note that the distances between adjacent sites differ; those distances reflect not only a measure of shared types, how similar two sites are, but, Dempsey and Baumhoff (1963:507) reasoned, also temporal differences. But the only way the latter could be so is if the rate of change was constant over the entire temporal span represented by the line. Drennan (1976:295) presented a figure similar to our Fig. 4.9, but he also correctly cautioned that the line represented " 'time' measured in terms of how much the ceramics have changed rather than time measured in actual years" and that the rate of change, measured against years, could be different at different positions along the line (Drennan, 1976:299).

A FINAL NOTE

Despite recognition of the fact that seriation provides "a more detailed picture of temporal change than stratigraphy and field observations alone" (Meighan, 1959:203), about 20 years after Kroeber invented frequency seriation it was used less and less as a dating method. This decrease in use occurred because archaeologists shifted from using stratigraphic excavation to confirm that an ordering produced by seriation was chronological to using excavation to construct a chronology of artifacts (Lyman and O'Brien, 1999; Lyman *et al.*, 1998). The invention of radiocarbon dating in the late 1940s exacerbated this trend, and occurrence and frequency seriation, though the subject of some discussion in the 1960s and early 1970s as a result of the increased availability of computers to do the sorting, never enjoyed a level of use such as they had in the 1920s and 1930s. Seriation is the only relative dating technique that allows time to be measured in anything approximating a continuum. As we will see in the next two chapters, stratigraphic superposition and typological cross dating tend to measure time discontinuously. Anyone interested in measuring time as a continuum must be cognizant then of the seriation method and the various techniques it comprises.

5 Superposition and Stratigraphy

Measuring Time Discontinuously

Ask any ten people you meet on a street corner to give you one or two sentences that characterize what archaeologists do, and the chances are excellent that all ten will include the clause, "they dig," or some variation thereof in their responses. There is no getting around the fact that digging, or as archaeologists prefer to say, "excavating," is central to the discipline. The reason for this is obvious: Artifacts are found on or in the ground, and in the case of the latter, the only means of recovering them is through excavation. Since the modern ground surface is only the latest in a long series of previous surfaces, it stands to reason that the bulk of the world's archaeological record is buried. Archaeologists perform myriad other tasks besides moving earth in the search for artifacts, ones that seem to occupy a large percentage of their time, such as analysis and report preparation; but if we trace our way back through those activities, we more than likely will find that the path leads back to one or more periods of excavation. Most archaeologists probably first entered the profession because they had a desire to find something, to experience the sense of discovery that comes from fieldwork, especially from excavation. We imagine that the majority of archaeologists stay in the discipline for that same reason. They have an intrinsic desire to know more about the past, and they do not mind putting up with the vagaries of nature—a broiling sun, torrential rains, mosquitoes, and the like—to satisfy that desire.

Archaeologists are trained in excavation technique almost from the first moment they express an interest in the discipline. It is not uncommon for colleges and universities to run weekend classes for undergraduates in how to excavate a site properly, and it is routine for research universities to hold 4- or 8-week field schools where both graduate and undergraduate students receive extensive training in excavation procedures. Nor is it out of the ordinary for universities to sponsor training exercises for interested nonprofessionals connected with state archaeological societies. Our experience has been that it is not difficult to train people in careful excavation technique. Most people, regardless of their level of experience, seem to take seriously the need to lay out excavation units carefully; to excavate in nice, neat units; and to keep separate artifacts from different levels or other segments of a site. They all have seen pictures of excavations on television and in magazines, and they understand that archaeology is, at a minimum, supposed to be an orderly affair.

Sometimes excavations are carried out in such orderly fashion that the entire exercise takes on an air of sanctity. There is a seldom-remarked rule that archaeologists are to be neat in how they excavate; it is based on the fear that a colleague might show up and cast a disapproving glance at a bowed-out sidewall in an excavation pit or an uneven floor of a unit. As British archaeologist Sir Mortimer Wheeler (1956:80) noted, "On approaching an excavation, the trained observer can at a glance evaluate its efficiency. It is an axiom that an untidy excavation is a bad one." Therefore, archaeologists keep straight walls in excavation units (Fig. 5.1), they pay attention to where artifacts came from within the site, both vertically and horizontally, and they take high-quality photographs and make detailed maps of the locations of the artifacts they find. Hopefully such care and precision will allow them to make an accurate reporting of what the site was like to those of us who were not there. Reporting is the easy part; interpreting the depositional history of the site is more difficult.

A number of excellent manuals cover the myriad details of how and how not to excavate an archaeological site, and some of them point out the complexities involved in interpreting the history of a site (Fig. 5.2). Our focus here is not on excavation technique but rather on one select aspect of excavation, namely, how can we make legitimate use of depositional history at a site to infer the passage of time? The answer to this question appears simple, an answer that one would assume all first-year archaeology students could answer correctly: You make predictions about the passage of time by correctly reading stratigraphic evidence—literally, evidence that tells us the actual sequence of events that took place at a site. As a shorthand device, we will call stratigraphic evidence *stratigraphy*, though that term in its pure form means the study of such evidence.

Archaeologists, like geologists, have long known that vertical sequences of sediments can be read in timelike fashion (Lyman and O'Brien, 1999; Praetzellis, 1993; Rowe, 1961). In short, everything else being equal, things deposited first are at the bottom of a column of sediments, and those deposited last are on top. Thus we have a measure of relative time; all we have to do is to make sure that everything really is equal—that nothing has happened to disrupt the correct order of whatever depositional units happen to appear at a site. This sounds simple enough in principle: If one excavates carefully and takes good notes and draws detailed maps of excavation walls, then it should be apparent if, say, a prehistoric pit penetrated a lower (and thus earlier) layer, thereby perhaps adding later artifacts to the earlier layer (Fig. 5.3). If we see this intrusion, then we can correct for it, thereby making everything "equal" again.

Stratigraphic interpretation has become so routine in archaeology that it is second nature to most of us. We know that older things are supposed to be at the bottom of a site and later things on top, so it seems as if there is no reason to belabor the point. And yet there is a danger that comes with familiarity of a subject, especially when we are so familiar with it that we do not think about it anymore.

Figure 5.1. Photograph of excavations of Mulberry Creek shell mound, Colbert County, Alabama, 1937 (from Webb and DeJarnette, 1942).

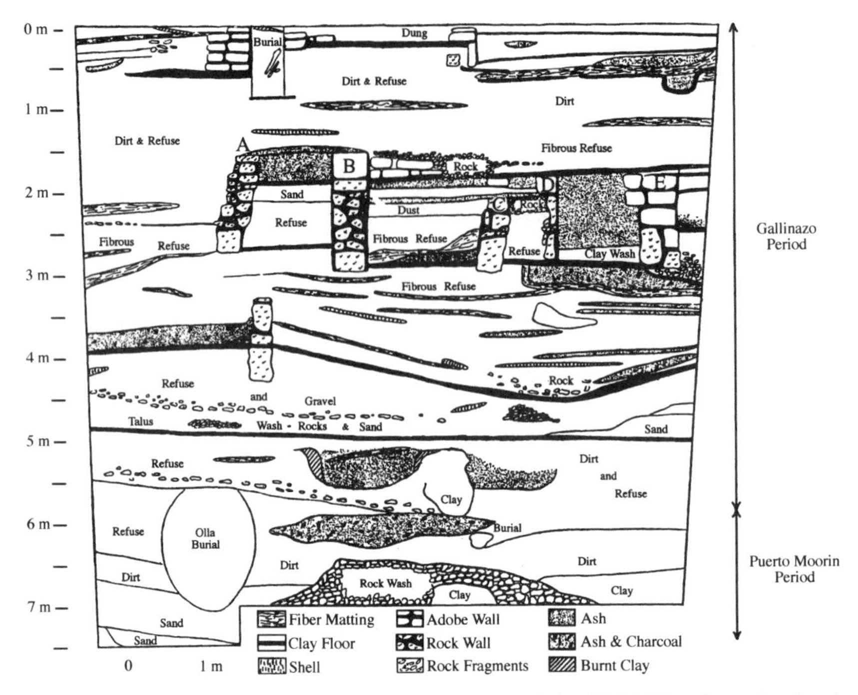

Figure 5.2. Profile of strata cut I at the Castillo de Tomaval site, Virú Valley, Peru, showing the complexity of superposed deposits consisting of various kinds of archaeological and nonarchaeological sediments (after Strong and Evans, 1952).

How many of us ever step back and ask ourselves how solid are the stratigraphic interpretations that we routinely make? How solid is our knowledge of stratigraphic principles? Do we really understand the difference between superposition and stratigraphy? Based on our experience (Lyman and O'Brien, 1999), there is room for serious concern. It appears from close inspection of the literature that there is considerable variation not only over what stratigraphy is but what stratigraphic excavation entails. Some modern textbooks do not help matters when they use such terms as stratigraphy and stratification interchangeably or fail to mention the relation of those terms to superposition. Nor do they do the discipline much service by leaving terms such as "stratum" undefined.

Part of the problem stems from a separation of today's archaeologist from the period when stratigraphic principles were openly discussed and debated, namely, the culture historical period of earlier this century (Lyman *et al.*, 1997b). Modern archaeologists tend to shy away from the "ancient writings" of their predecessors, figuring that there is not much of relevance in them, but with regard to superposi-

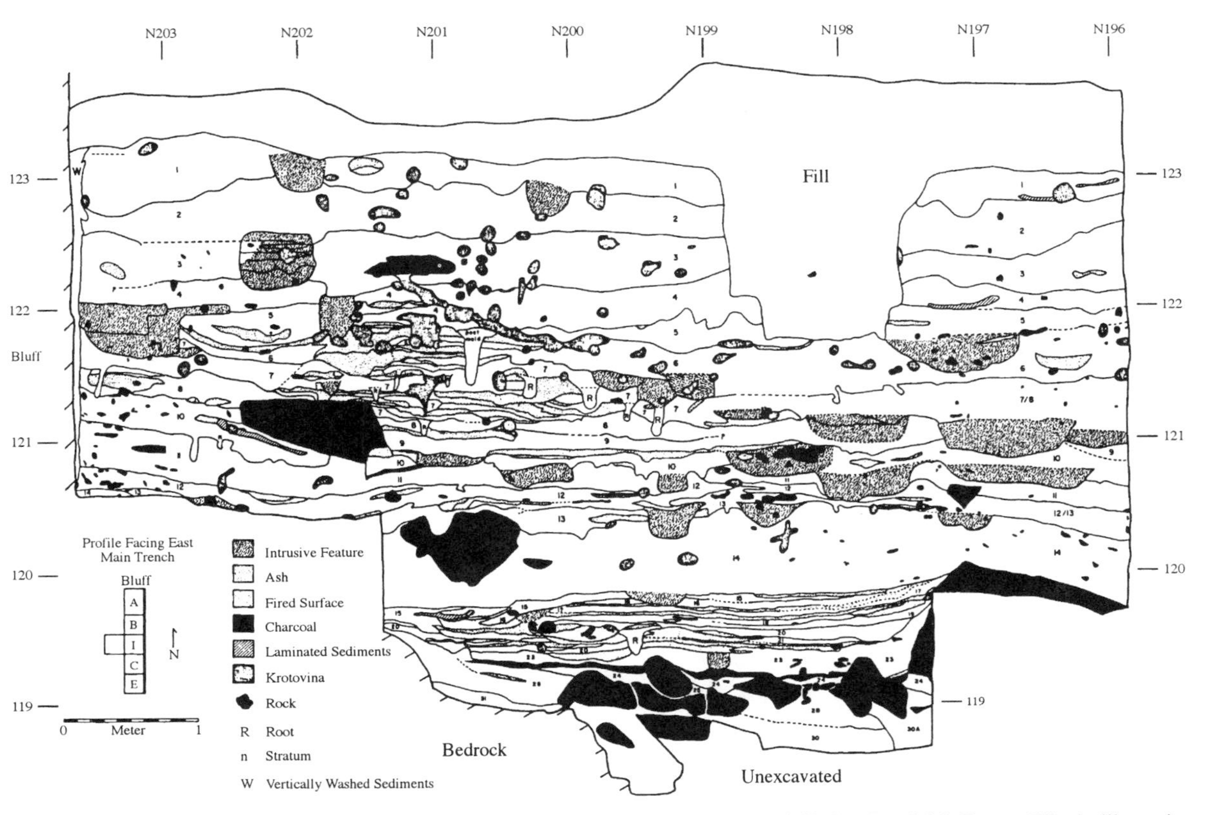

Figure 5.3. North–south profile of 1984 main trench in the main shelter area at Modoc Rock Shelter, Randolph County, Illinois, illustrating the confusing nature of the deposits in and disturbances to strata caused by the excavation of pits prehistorically and by animal burrowing (labeled on the profile as "krotovinas") (after Ahler, 1993).

tion and stratigraphy this is a mistake (Lyman and O'Brien, 1999). Numerous treatments from the first half of this century were models of stratigraphic analysis, clearly laying out in plain language not only the problems involved in making sound interpretations but also the principles that underlie the interpretations. This is not an indictment of modern archaeology, nor is it a veiled attempt to promote the work of culture historians as being flawless. It is fair to say that numerous archaeologists working prior to, say, 1960, failed to realize the problems they were creating by ignoring not so much the mechanics of site formation, which became a growing concern in the 1970s with the work of Michael B. Schiffer (e.g., 1972, 1987) and others, but rather the basic principles of how to use stratigraphic evidence to mark the passage of time. Simply because someone was involved in doing culture history did not mean that he or she had received the wisdom of an A. V. Kidder or a Nels C. Nelson when it came to telling time by means of stratigraphic observation. And even eminent archaeologists such as Kidder and Nelson often ended up making the fatal mistake of equating strata with cultures.

STRATA, STRATIGRAPHY, AND SUPERPOSITION

We need first to examine what we mean by certain terms. The most appropriate place to begin is with the term *stratum*, which we define as a three-dimensional unit of sediment of any origin that represents a depositional event and is distinguishable from other such units (Lyman and O'Brien, 1999; Stein, 1990). Importantly, this definition emphasizes that a stratum arises from a single depositional event; inclusion of the words "depositional event" denotes that all of the sedimentary particles, including artifacts, comprising a stratum were deposited more or less contemporaneously.

Our use is much more in line with how geologists use the term stratum (e.g., Stein, 1990) than how archaeologists use it. In archaeology strata are often termed "levels" or "layers," and vexingly they often are referred to as "cultures." Note how Robert Ehrich (1950:473) interchanged those terms: "Each layer in a stratified site is tentatively called a separate culture, at least until it can be related to earlier or later levels within the same excavation or to layers with a similar content in other sites or areas." But strata are not cultures, nor are they in and of themselves necessarily indicative of anything cultural. They are units that comprise individual depositional events that may or may not be the result of human activity. Where units are superposed, or layered, the principle (some would say "law") of superposition states that the one on the bottom was deposited first and the one on top deposited last. Everything in between falls in relative chronological (depositional) order. This principle has been around a long time, with credit for its formulation usually given to Nicolaus Steno, a nineteenth-century physician in the court of Ferdinand II, Grand Duke of Tuscany.

Note that the principle of superposition prescribes the depositional order of strata and is an ordinal scale measurement of time. Does the principle apply only to naturally deposited sediments? No. Recall that we defined a depositional event, which results in the creation of strata, as comprising sedimentary particles, including artifacts, that were deposited at approximately the same time. The principle applies equally to both artifacts and natural sediments. Everything in a stratum, from the tiniest of clay particles to the largest of objects including artifacts, predates everything in the stratum above in terms of when it was deposited. We emphasize that the principle of superposition says nothing about the ages of the sediments—when they were formed—in each stratum relative to the ages of the sediments in other strata above or below it. It says only that one set of sediments was *deposited* at a particular locality before or after another set was.

Conflating the age of sediments with the relative positioning of strata is the single-most common stratigraphical mistake made by archaeologists, and it can be a fatal one in terms of chronological ordering. The misconception that the two are always related was and still is perpetuated in textbooks, as exemplified by the following statement:

> [T]he layers deposited first must be the oldest, and the levels above must come later in time. This explains why archaeologists are so careful to record the depth at which objects are found—the depth usually has a bearing on the age of the find, at least in relating it to other finds from the same site. (Meighan, 1966:26)

The first sentence concerns the age of the depositional event, but the second sentence confuses the age of that event with the age of the sedimentary particles. Sadly, such errors are still made, as evidenced by this statement: "Relative chronology is based on the simple stratigraphic principle that older materials will be found lower in an archaeological deposit than newer materials—the law of superposition" (Michaels, 1996:168). Such statements fail to recognize that superposition is an indirect dating technique when applied to artifacts within strata. It is indirect because the age of the artifacts is inferred from their vertical positions relative to one another; their positions are extrinsic properties. Another way to say this is that the target event is the age of the artifact's creation, whereas the dated event is the age of the depositional event.

Suppose we have a sedimentary column, regardless of whether it contains artifacts, and in it we can clearly identify five strata on the basis of various sedimentary characteristics: grain size, texture, color, and so on. We know that the stratum at the bottom of the column predates all the other strata and that the one on top postdates all the others in terms of when they were deposited. But what about the ages of the sediments comprising the five strata? Why should they reflect any particular order? Perhaps the parent material that produced the sediments in the lowermost stratum actually is younger than the parent material that produced sediments in the stratum just above the lowermost one. Maybe the parent material that produced the sediments in the uppermost stratum actually is older than all the other parent material that contributed sediments to the column.

Is this a hypothetical example, or does it happen that way in the real world? The answer is that it *does* happen and rather frequently at that. Suppose that the lowermost stratum in our imaginary column contains sand-sized particles deposited by a river that had cut through decomposing Cretaceous age limestone that had formed on an ocean floor 80 million years ago. The flow rate of the river was such that it carried the particles in suspension for several miles before the rate lessened, at which point the particles were deposited. Then, there was no more fluvial deposition at that locality for over a century, but during that time surrounding trees and bushes made an annual contribution of leaves and other organic debris that decomposed in place, forming a thick organic stratum. At the end of the century the river again deposited sediments at the locality in question, only this time the heavy fraction was derived from limestone that had been deposited approximately 235 million years ago, near the end of the Permian period. If we excavated the sedimentary column just after the river made its second delivery, we would have three strata, each distinct from the others and all in chronological order relative to when they were deposited, but the ages of the sediments themselves, from bottom to top, would be 80 million years old, roughly 50 years old, and 235 million years old. Not exactly a chronological arrangement in terms of the ages of the sediments.

Now suppose that as the sedimentary column is accreting, say around 4000 BC, a group of humans camps on the spot and leaves behind some tools, which become buried as more and more sediments are deposited as a result of flooding from the river. Then the area becomes drier, which causes deforestation of the hillside just above where the column is accreting. The hillside erodes, which causes tools left around 7000 BC by a group of hillside dwellers to wash downslope, where they lay on the surface of the accreting column. Then, more sediments are carried in to form the uppermost stratum. Thus we have a nice, orderly progression of sediment buildup. We also have a nice, orderly progression of strata, from the oldest (first deposited) on the bottom to the youngest (last deposited) on top. In addition, we have 9000-year-old artifacts superposed on 6000-year-old artifacts. Archaeologists, if and when they recognize it, refer to this as reversed or mixed stratigraphy, but there is nothing "reversed" or "mixed" about it. Everything is just as nature left it. The reversed-stratigraphy argument rests on the assumption that we know what the proper sequence of artifacts is. Whether we do or not, our point is, why should we place our faith in stratigraphic positioning without asking ourselves if there is a "systematic relationship between the contents and layer deposition" (Dunnell, 1981:75)?

The work of the archaeologist, like that of the geologist, is to analyze the superposed sediments and to determine when strata were deposited as well as when the sediments were formed. Archaeologists are interested in culturally derived sediments—artifacts—but they realize that the artifacts usually occur within noncultural, or natural, sediments. The nature of the latter often are an important source of information relative to the nature of the former. For instance, in our example above, close examination of the stratum containing the hillslope-derived artifacts might reveal that the sediments within the stratum are identical to those on the hillslope. From that evidence we could propose that the artifacts mixed in with those natural sediments washed downslope with them. What we are really trying to understand are site formation processes (Schiffer, 1987). They must be understood before we can use artifacts from strata to measure the passage of time.

Not all individuals will identify deposits, strata, or so-called "natural layers" in exactly the same way because they use different criteria to distinguish among them. This means that the temporal boundaries and durations of the depositional event represented by such a unit will vary between investigators (Stein, 1990). Despite these differences, most archaeologists would nonetheless argue that there is considerable merit in keeping the artifacts found in natural depositional units separate for purposes of analysis (Deetz, 1967; Praetzellis, 1993; Rathje and Schiffer, 1982). The reason for doing so is obvious: Artifacts within a stratum are associated stratigraphically (were deposited at more or less the same time), and thus are potentially of approximately the same age (were used at the same time). One might infer that stratigraphically associated artifacts are more or less contemporary in terms of when they were made and used.

Do not be misled, however. The principle of stratigraphic association is only as good as the criteria used to segregate strata. No principle can save sloppy stratigraphic analysis. We need to remember that the principle is an inference, a point made clear in numerous studies of site formation (e.g., Schiffer, 1987). Take our example of artifacts washing down the hillslope above where our sedimentary column is accreting. Suppose erosion was severe enough that tools left even higher up the slope by a group around 4000 BC were transported down to where they joined those left around 7000 BC, and together they were transported down to the bottom, where they were incorporated into the same stratum. In such a case the local sequence would need to be well known to allow such mixing to be spotted.

Stratigraphic Excavation

Archaeologists excavate, but unless they make use of strata as artifact collection units, they are not excavating stratigraphically. In other words, the excavation has to result in sets of artifacts being recovered from vertically distinct units. But what does vertically distinct units mean? We have indicated that such strata are visibly distinct, but there is no reason why we cannot create strata ourselves. Nels Nelson, for example, simply divided the midden at Pueblo San Cristobal into 1-foot-thick units and excavated each in turn (Fig. 5.4). Some individuals refer to arbitrary, or metric, excavation levels as "strata" (e.g., Nesbitt, 1938), which they are, and others state or imply that the use of arbitrary levels as

Figure 5.4. Stratigraphic cut made by Nels C. Nelson through midden deposit at San Cristobal, New Mexico (from Wissler, 1921) (photo courtesy American Museum of Natural History).

artifact-collection units comprises stratigraphic excavation (e.g., Browman and Givens, 1996; Deetz, 1967; Givens, 1996; Heizer, 1959; Taylor, 1954; Thomas, 1998; Willey and Sabloff, 1993), which it does. Gordon Willey (1953:365, 367) labeled such excavation "continuous stratigraphy" and "vertical continuous stratigraphy" to underscore the absence of natural breaks in a sediment column. Certainly the boundaries of arbitrary units that are vertically distinct from one another define a period of deposition, just as the boundaries of "natural" units do, though the former may represent only a fraction of the "natural" depositional event represented by the stratum or deposit from which they are derived.

We define stratigraphic excavation as removing artifacts and sediments from vertically discrete three-dimensional units of deposition and keeping those artifacts in sets based on their distinct vertical recovery proveniences for the purpose of measuring time. The vertical boundaries of the spatial units from which artifacts are collected can be based on geological criteria such as sediment color or texture, or on metric (arbitrary) criteria such as elevation. The relative vertical recovery provenience of the artifacts is what is critical, and it stems from the basic notion of superposition. Vertical provenience may be relative to an arbitrarily located horizontal datum plane, and thus the archaeologist knows that one artifact was found at a higher elevation than another or that two artifacts were found at the same elevation, but the archaeologist may not know how those vertical locations relate to geological depositional events. Or, vertical recovery proveniences may correspond to some "natural" depositional units or strata. Artifacts recovered from one (or several contiguous) vertical location(s), as denoted by arbitrary levels or strata, are inferred to be approximately contemporaneous based on the principle of association; artifacts recovered from superposed strata or arbitrary levels are inferred to be of different ages based on the principle of superposition.

STRATIGRAPHIC EXCAVATION IN HISTORICAL CONTEXT[1]

The careful stratigraphic work that archaeologists do today has its roots deep in Americanist archaeology. Received wisdom is that sometime in the second decade of the twentieth century archaeologists suddenly and without much foundation began excavating stratigraphically. Although it is not explicit in most discussions of the history of the "stratigraphic revolution" (e.g., Browman and Givens, 1996; Willey and Sabloff, 1993; Woodbury, 1960a), it appears that what this term means is that when Americanist archaeologists suddenly became interested in time, excavation techniques were changed as a result. In short, the beginnings of stratigraphic excavation are thought to mark the stratigraphic revolution.

[1]Some of the following discussion is adapted from Lyman *et al.*, (1997b) and Lyman and O'Brien (1999).

We believe, however, that historians of archaeology have overlooked the fact that stratigraphic excavation has a much deeper past in Americanist archaeology. We think the problem resides in how stratigraphic excavation is being defined. Our definition, remember, states that stratigraphic excavation entails removing artifacts and sediments from vertically discrete three-dimensional units of deposition and keeping those artifacts in sets based on their distinct vertical recovery proveniences for the purpose of measuring time. And that is all the definition says. Note that it says nothing about how a site is excavated to make the strata visible. One might peel the strata back horizontally one at a time, what Gordon Willey in his unpublished master's thesis (Willey, 1936:56) termed the "pure 'onion peel'" technique (Fig. 5.5), or slice off vertical faces, what Willey (1936:58) described as "slicing like a loaf of bread." But simply because one does not excavate in one of those two manners does not indicate an absence of stratigraphic excavation, if what is meant by this term is the use of discrete and distinct depositional units as artifact recovery units.

David Browman and Douglas Givens (1996:81) distinguish between "post facto stratigraphic observation" and "actual stratigraphic excavation." They equate the former with the identification of "archaeological strata ... in the walls of

Figure 5.5. Photograph of excavations of the Bluff Creek mound, Lauderdale County, Alabama, 1937, showing the combined block-outline and onion-peel techniques of excavation (from Webb and DeJarnette, 1942).

trenches excavated as single [vertical] units" and the latter with the identification of "archaeological strata [that are] microstrata of geological units" and the employment of those strata as "data recovery units" (Browman and Givens, 1996:80). This distinction is logical enough, but it is so simple that it obscures the beauty of what actually happened. Reliance on such a dichotomy, although technically sound, forces us to focus on excavation technique and not on the central issue: whether time was being measured using artifacts recovered from strata. And, when we make the shift back to that question, we find that many prehistorians prior to the second decade of the twentieth century were doing exactly that. To us, it makes no difference if they used onion peel, bread loaf, or some other excavation technique.

Prior to the so-called stratigraphic revolution, shallow time was of great interest and resulted in much excavation that by our definition was stratigraphic. Americanists prior to the second decade of the twentieth century were well aware not only of the principles of superposition and association but also of how to use them to infer time (Lyman and O'Brien, 1999; Lyman *et al.*, 1997b; Rowe, 1961). Their awareness stemmed from the fact that, whether they were interested in shallow or deep time (or both), and most often it was the former, they knew that to measure either sort of time artifacts had to be collected from vertically discrete units. Where then did stratigraphic excavation come from?

Early Stratigraphic Excavation

The historical problem of identifying the first stratigraphic excavation in the Americas may never be solved, but some might refer to Thomas Jefferson's 1784 excavation of a trench through one of the earthen mounds on his property in Virginia. He subsequently made notes on the stratigraphic relation of layers of earth and human bones in the mound and remarked on the chronological implications of the layering:

> At the bottom [of the mound], that is, on the level of the circumjacent plane, I found bones; above these a few stones ... then a large interval of earth, then a stratum of bones, and so on. At one end of the section were four strata of bones plainly distinguishable; at the other, three; the strata of one part not ranging with those in another.... Appearances certainly indicate that [the mound] has derived both origin and growth from the accustomary collection of bones, and deposition of them together; that the first collection had been deposited on the common surface of the earth, a few stones put over it, and then a covering of earth, that the second had been laid on this, had covered more or less of it in proportion to the number of bones, and was then also covered with earth; and so on. (Jefferson, 1801:141–142)

Jefferson's work was later said to have anticipated modern archaeological methods by a century and to have been undertaken "to resolve an archaeological problem" (Lehmann-Hartleben, 1943:163), but it had no visible impact on Americanist archaeology. As Mortimer Wheeler (1956:59) later observed, "Unfortunately, this seed of a new scientific skill fell upon infertile soil. For a century after Jefferson, mass-excavation remained the rule of the day." It was 75 years later, and then in Great Britain, that Jefferson's approach was reinvented with effect, having been used sporadically in Europe since the late eighteenth century (Van Riper, 1993). The excavation of Brixham Cave by prominent British geologists and paleontologists in 1858 focused explicitly on stratigraphic context. That work, which was based on Danish prehistorian J. J. A. Worsaae's (1849) principle of association, resulted not only in the recognition that human antiquity was deeper than indicated by biblical history but also in archaeology becoming a legitimate scholarly pursuit in the Old World (Grayson, 1983; Van Riper, 1993).

Worsaae's associative work was designed to demonstrate that the sequence of stages earlier proposed by Christian J. Thomsen (1848), that stone tools appeared first in the archaeological record, followed by those of bronze and finally by those of iron, was correct. It is important to note that Worsaae's strongest evidence came not from stratigraphy but from the fact that tools of different material occurred in different kinds of graves and alongside different artifacts:

> [F]or as to the graves themselves we know that, generally speaking, they contain both the bones of the dead, and many of their weapons, implements, and trinkets, which were buried with them. Here we may therefore, in general, expect to find those objects together which were originally used at the same period. (Worsaae 1849:77)

Although superposition was not integral to Worsaae's argument, there is no question that prehistorians of the time were making stratigraphic observations, as is seen in Fig. 5.6, which appeared alongside Worsaae's discussion.

The history of Americanist archaeology makes it clear that the first archaeological applications of stratigraphic principles were by natural historians and those with parallel interests; there were no trained archaeologists in the United States during the period 1860–1900. Physician Jeffries Wyman, who served as the first curator of the Peabody Museum (Harvard), was one such naturalist. In the 1860s he examined shell mounds in New England, observing that "A section through the heap at its thickest part showed that it belonged to two different periods, indicated by two distinct layers of shells" separated by "a layer of dark vegetable mould, mixed with earth and gravel" (Wyman, 1868:564). These observations prompted Wyman to suggest that a sequence of two occupations separated by a period of abandonment was represented by the strata. A few years later, Wyman (1875:11) indicated that the absence of artifacts of Euro-American manufacture from shell mounds in Florida suggested that the mounds dated "before the white man landed on the shores of Florida." Cyrus Thomas of the Bureau of American Ethnology would make a similar case a few years later, only this time using the presence of European items to argue that some of the mounds in the Southeast dated to the period of European exploration.

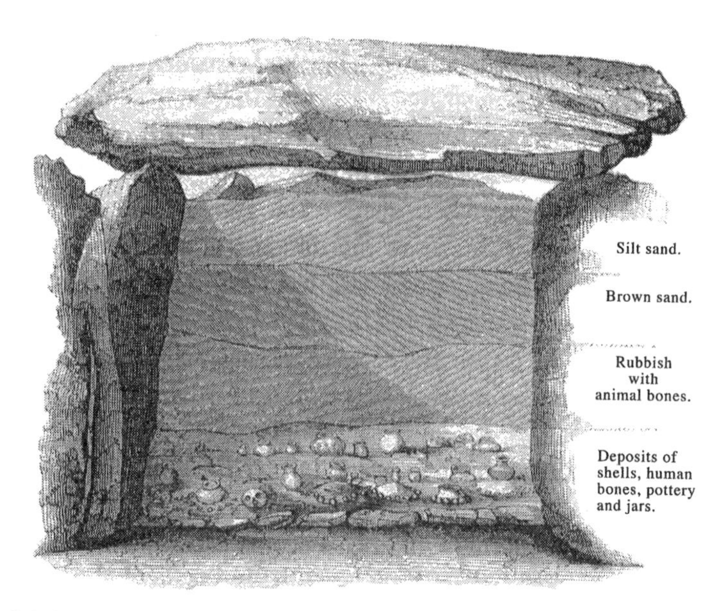

Figure 5.6. Stratigraphic section through a stone grave on the Isle of Guernsey showing different depositional strata (after Worsaae, 1849).

More or less at the same time Wyman was working in Florida, William Healey Dall, a conchologist, was excavating shell mounds in the Aleutian Islands (Dall, 1877). He not only recognized distinct strata in the mounds, but he excavated them in such a manner that he was able to record the stratigraphic provenience of many of the artifacts he recovered. He also was able to characterize, on the basis of food remains and artifacts, three temporal periods: an early hunting period, a middle fishing period, and a late littoral period. He referred to these not only as temporal units but also as evolutionary stages. It was in the same year that Dall's study appeared that Lewis Henry Morgan published *Ancient Society*, in which he outlined his unilinear evolutionary scheme for classifying societies (Morgan, 1877). Willey and Sabloff (1993:62) suggest Dall's work "had no impact at all on other archaeologists of the time," perhaps because of the "remoteness of the Aleutians" and because he did no follow-up excavations. More likely, it was because he was unable to overcome the general consensus not only that there was a very shallow time depth to the occupation of North America but, more importantly, that pre-AD 1500 cultures were similar to or the same as historically documented ones (Meltzer, 1983).

S. T. Walker (1883) did not excavate any shell mounds in Florida, but he closely examined the stratigraphy of those that had been exposed by construction-related activities, noting a three-stage sequence in the development of pottery. He characterized the earliest pottery "style" (Walker, 1883:679) as large, rude, heavy, and destitute of ornament; middle-period pottery as having thinner walls and some modification of the surface; and late-period pottery as being thinner because of the "employment of better materials." It also was "beautifully ornamented." Perhaps, then, an artifact's form, for pottery, at least, could be used to tell time, but as with those by Jeffries and Dall, Walker's report had little or no apparent impact on Americanist archaeology. Changes in artifact form were meaningless when viewed against the assumption of a shallow past.

Charles Peabody of Phillips Academy in Andover, Massachusetts, excavated Edwards Mound in Mississippi, in 1901–1902 (Peabody, 1904), noting that the stratigraphy indicated there were two periods of mound construction and use, and he reported differences between the artifacts and burials associated with the two periods. It was, however, Peabody's work in 1903 at Jacobs Cavern in McDonald County, Missouri, which he conducted with Warren Moorehead, that was based on a sophisticated excavation strategy. Peabody and Moorehead gridded the cave floor into a series of 1-meter-square units (Fig. 5.7) and then removed the deposits

> in order, front to back, using the lines of stakes as coördinates to determine the position of any objects found. The linear distances of 1 m. were numbered from northwest to southeast in Arabic numerals from 1 to 21, and lettered from southwest to northeast from A to Q. (Peabody and Moorehead, 1904:13)

Three vertical levels were recognized in the deposit, though it is clear from reading their account that they did not pay too much attention to separating artifacts by level as they excavated.

Peabody regularly used a horizontal grid to guide his excavation, and he consistently recorded the vertical and horizontal provenience of artifacts, burials, and features (e.g., Peabody, 1908, 1910, 1913). He was well aware that more deeply buried materials could be inferred to be older than materials located near the surface, and given how he excavated, he generally knew which strata produced which artifacts. However, he seldom spoke of temporal differences as measured by the relations of the vertical positions from which his artifacts were collected.

Mark Harrington's (1909a,b) work in New York in 1900–1901 was similar to Peabody's in that Harrington not only made post facto stratigraphic observations but also used stratigraphic excavation. He first excavated several trenches in a rock shelter, noting there were three strata, the middle one containing no artifacts. He indicated which artifacts came from which stratum and presented two drawings of "vertical sections" of the stratigraphy (Harrington, 1909b:126–129). He then wrote:

> When the [excavations had] been completed, it was thought that everything of value had been found and removed from the cave; but on further deliberation, taking into consideration the darkness of the cave and the blackness and stickiness of the cave dirt, it

Figure 5.7. Photograph of Charles Peabody and Warren K. Moorehead's excavation of Jacobs Cavern, McDonald County, Missouri, 1903 (from Peabody and Moorehead, 1904) (courtesy Phillips Academy, Andover, Mass.).

> was thought best to sift the entire contents. The results were surprising. The earth had all been carefully trowelled over, then thrown with a shovel so that it could be watched—but a great number of things had been overlooked, as the subsequent sifting showed. Of course all data as to depth and position have been lost, yet the specimens are valuable as having come from the cave. (Harrington, 1909b:128)

Harrington was well aware of the temporal significance of superposed strata and the provenience of artifacts in those strata, and during his excavation he used depositional units as artifact collection units. He did not discuss, however, the cultural or temporal differences between his two superposed collections, except to note that the stratigraphically uppermost level contained pottery, whereas the lowermost did not. Thus, his excavation was not geared toward measuring time.

Harrington's (1924:235) description of how he excavated a site in New York in 1902 is the most revealing statement we have encountered concerning whether Americanist archaeologists were excavating stratigraphically prior to 1912. He indicated that after digging "test holes [to determine] the depth and richness of the deposit [he chose the parts of the site that] warranted more thorough excavation." Then, he exposed a vertical face "at the edge of the [site] deposit." This vertical face was extended down through the "village layer" or "the accumulated refuse of the Indian Village," which included various depositional units such as layers of shell fish, ash, and black sediment, into the underlying sterile stratum: "A trench of this kind was carried forward by carefully digging down the front with a trowel, searching the soil for relics, then, with a shovel, throwing the loose earth thus accumulated back out of the way into the part already dug over, so as to expose a new front." Once a trench was completed, "another trench was run parallel and adjacent to the first on its richest side, and so on, until the investigator was satisfied that he had covered the entire deposit, or at least as much as his purpose required" (Harrington, 1924:235). As a result, Harrington noted that different artifacts came from different strata. For example, in the deepest stratum he found "stemmed arrow points and crude crumbling pottery of a somewhat more archaic character than most of the specimens" recovered from higher strata (Harrington, 1924:245). He also noted that the archaic artifacts were stratigraphically "below the village layer" (Harrington, 1924:283).

Harrington's discussion underscores the fact that although Americanists did not peel back strata (or arbitrary levels) one at a time (the onion-peel technique) prior to 1912, many of them did employ strata as artifact recovery units, often by using the bread-loaf technique. It also underscores the fact that it actually matters little how the sediment is removed from the hole one is digging, as long as the stratigraphic proveniences of the artifacts are recorded. And Harrington used artifact provenience to measure the passage of time.

Peabody and Harrington were not the only archaeologists employing strata as collection units during the pre-1912 period, nor were they the only ones to infer that superposed artifacts and/or deposits measured time (see Lyman and O'Brien, 1999, for more details). Almost everyone made this assumption. But the point is that none of these archaeologists was overly concerned with chronological matters, even though the artifacts they collected were known to fall in different time periods, given their provenience in different strata. The question is, why were they not concerned? The work of Fred H. Sterns, who in 1915 published an article entitled "A Stratification of Cultures in Nebraska," contains the answer to this question. Sterns observed that "the proof of [cultural] sequences must be grounded on stratigraphic evidence, and stratified sites have been very rare [in North America]. Hence such a site has a high scarcity value and warrants special study even though it be otherwise of minor importance" (Sterns, 1915:121). Sterns, like many of his contemporaries, did not devote much time to chronological matters. They did not because many of them, like Sterns (1915:125), observed that "An important fact arguing against any great difference in time between the upper and lower ash-beds is that the pottery and the flint and bone implements found in these two sets of fireplaces show absolutely no difference in type." This is simply a rewording of Franz Boas's (1902) earlier statement that there were no significant differences between historically documented cultures and prehistoric ones due to the shallow time depth then ascribed to the American archaeological record. Thus no one consistently or rigorously asked chronological questions, despite the fact that many of them collected artifacts from distinct depositional units.

On the Eve of the "Revolution"

Late in the first decade of the twentieth century, Mesoamericanist Zelia Nuttall was sufficiently familiar with the archaeological record of the Valley of Mexico to suspect that different kinds of clay figurines and pottery represented temporally distinct cultures. She therefore asked Mexican workmen to collect artifacts from beneath the Pedregal lava flow, artifacts that because of their position clearly "antedated any Aztec remains" (Nuttall, 1926:246), which had been found above the lava flow. At about the same time, Manuel Gamio, a student of Boas's, "recognized [what he believed was] a succession of various styles of figurines and pottery" (Vaillant, 1935a:289). How could the suspected chronological relations of these various materials be validated empirically?

As is well documented by Browman and Givens (1996:88–89), Boas was aware of the temporal significance of superposed archaeological materials, and he wanted to use stratigraphy not only to address chronological questions but also to wake his archaeological colleagues up to the fact that stratigraphic observation was a means to measure the passage of time. This was not a surprising position for Boas to take, given that he had served in several positions under Frederic W. Putnam, who was a supporter of a deep antiquity to the occupation of North America. We suspect that all Boas was really trying to do was to integrate archaeology with anthropology and to flesh out the results of one with those of the other. On the advice of Boas, Gamio (1913) excavated several deep trenches at the site of Atzcapotzalco in Mexico City in 1911–1912, primarily to test the validity of the sequence of Mexican pre-Columbian cultures then known as Tipo del Cerro (Archaic), Teotihuacán, and Aztec. His published stratigraphic profile is shown in Fig. 5.8.

Gamio (1913) ultimately identified the superposed relations of the three types of remains. From a historical perspective what is important is that Gamio's approach to the chronological problem concerned a suspected local sequence and the fact that superposition was used to confirm rather than to create that sequence. Also important is the fact that Gamio's excavation showed that the pottery types occupied nearly unique stratigraphic positions with minimal overlap. This is similar to earlier discussions of superposed artifact collections in which different kinds of artifacts were shown to occupy different strata (e.g., Boas, 1900; Harrington, 1909b). Gamio (1917) (see also Gamio, 1924) later suggested the waxing and waning of a type indicated the growth and decline of a population, thereby maintaining his distinction of three discrete cultural populations that were responsible for the superposed remains.

Gamio has achieved a place in the annals of archaeological stratigraphy (Adams, 1960), but it pales in comparison to that of a group of Southwesternists, whom we discuss below. Part of the reason for his nonstarring role probably has to do with the fact that Gamio published much of his work in Mexico, and thus was not read as widely as American prehistorians were. This is unfortunate, because we can see in Gamio's work a clear statement about the importance of superposed archaeological remains to the study of cultural change over time. Gamio excavated an extraordinarily valuable set of chronological markers, and they became excellent index fossils that were used across the Valley of Mexico in later years (Chapter 6).

The Real Revolution

There were four major players in what we see as the revolution of the teens: Nels Nelson, whose typological work we discussed in Chapter 2; A. V. Kidder and A. L. Kroeber, whose seriations we discussed in Chapters 3 and 4, respectively; and Leslie Spier. Each of them brought something different to the revolution, but each also built on what one or more of the others had just done. We exclude Kroeber from discussion because he did no excavation. The primary location of the revolution was the American Southwest, a vast region that had witnessed decades of excavation long before our players arrived there. Eastern institutions, including the US National Museum and the Bureau of American Ethnology, had put workers in the Southwest at an early date, but it was private expeditions, often financed by wealthy patrons, that carried out the bulk of the work. Much of it was little more than wholesale looting, with little interest paid to such things as chronology. The individuals we highlight here would soon change that.

Nels Nelson

Nelson's first major piece of fieldwork, which took place while he was a student at the University of California, was the excavation of the Emeryville

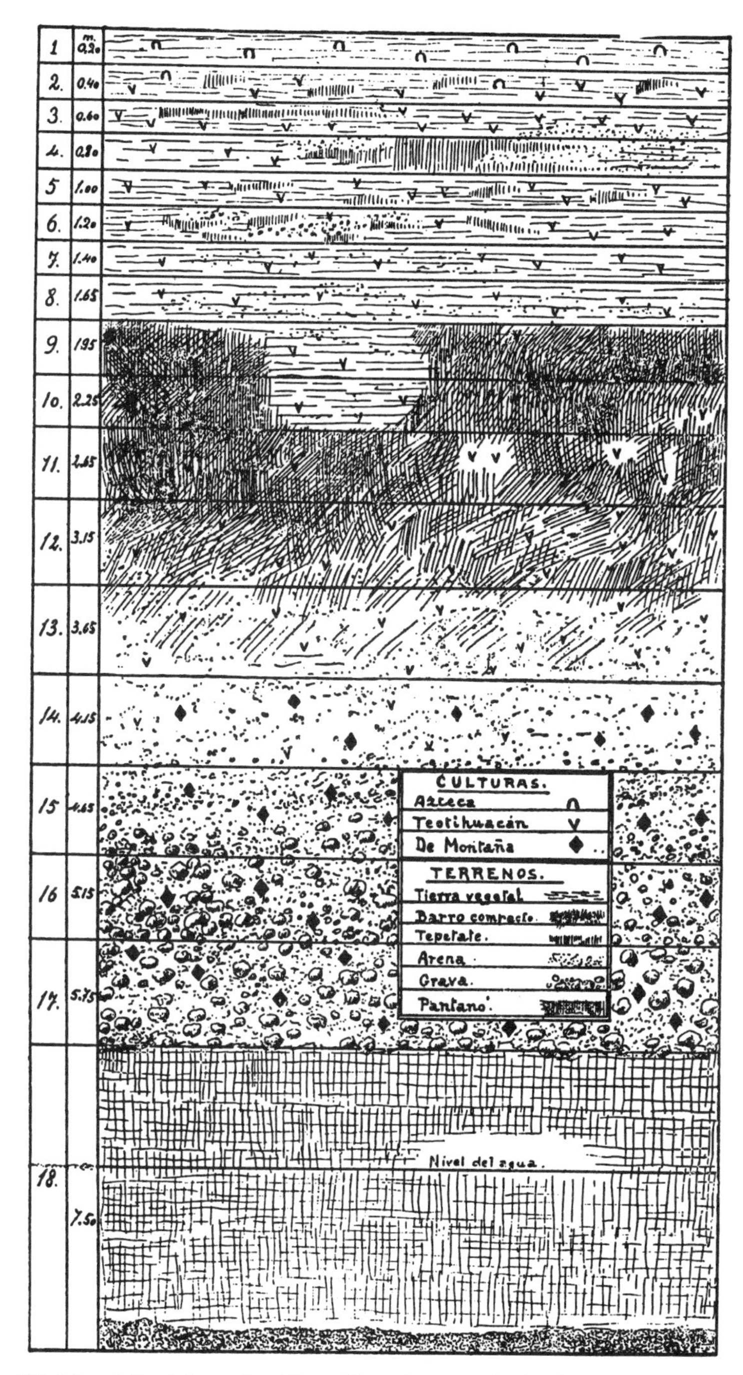

Figure 5.8. Manuel Gamio's stratigraphic profile at Atzcapotzalco in Mexico City. Note the varied thicknesses of the arbitrary levels and the vertical distribution of different artifact forms, which Gamio assigned to one of three cultures (inverted U, V, and diamond symbols) (after Gamio, 1913).

Shellmound in San Francisco Bay during May and June 1906. He is not well known by historians for his work there because until recently his report was unpublished (Nelson 1906). Nelson is better known (Woodbury, 1960a) for his excavation of Ellis Landing shell midden, also in San Francisco Bay, during August 1906 (Nelson, 1910). Nelson apparently was influenced by Max Uhle's slightly earlier work at Emeryville (Uhle, 1907), because at both Emeryville and Ellis Landing he recorded observations on stratigraphic positioning and site structure.

At Emeryville, Nelson (1906:5) noted that "Particular pains were taken to guard against objects of any kind falling accidentally to the working level from the walls of the [excavated] shaft and the possibility of such objects being located at a lower level than that at which they actually occurred." He used a "weighted tape [that could be] dropped [from an arbitrary datum plane] to any desired point in the shaft for vertical distances" (Nelson, 1906:4). Stratigraphic excavation was, in this case, good field technique. But also in this case, Nelson's care allowed him to compile a table showing the frequencies of artifacts and faunal remains of various types in each of 11 "strata." After excavating Ellis Landing, Nelson (1910:374) reported that "there are no well-defined strata of raw and calcined materials such as marked the upper part of the Emeryville mound [and] bedding planes [are] readily distinguishable only at some few points." Nelson also discussed his use of what sounds like the bread loaf excavation technique at Ellis Landing. In addition, when working below the water table, Nelson clearly excavated stratigraphically: "Under the circumstances, the dirt could not be carefully looked over at the time of removal from the [vertical] shaft [excavated beneath the water table], and was therefore laid out on the surface according to horizons [strata] and later thoroughly examined" (Nelson, 1910:373). This most definitely was stratigraphic excavation in that strata were employed as artifact collection units.

At Emeryville, Nelson (1906:11) observed that

> though the types of artifacts extracted from the shaft differ in some respects, the difference is not absolute, and the quality of workmanship is not so widely different as might be reasonably expected, considering the great period of time involved.... Dr. Uhle [1907] clearly found conditions quite different on the west side of the mound there being a striking gradation in the kind and quality of implements used.

Later, Nelson, being well aware of his mentor Kroeber's (1909) earlier salvo across Uhle's bow when Uhle (1907) had tried to talk about superposed artifacts and the passage of time, simply did not ask chronological questions at Ellis Landing, despite his estimate that the shell mound had formed over some 3500 years. Further, Uhle and Nelson, as well as Kroeber, were categorizing artifacts in technological and functional terms which, as we noted in Chapter 2, are not often useful for measuring time.

What Nelson is most remembered for is not his California work but rather his excavations in the Galisteo Basin of New Mexico, which he began in 1911, shortly after being hired at the American Museum of Natural History. Nelson's first report on his work in the Galisteo Basin documents that he was there for a very specific reason:

> It is felt that many problems relating to the origin and distribution of peoples and to cultural traits now observable in the Southwest cannot be solved in their entirety by the examination of present-day conditions or even by consulting Spanish documentary history, which though it takes us back nearly four hundred years and is reasonably accurate, shows us little more than the last phase of development within this most interesting ethnographic division of the United States. By a tolerably exhaustive study of the thousands of ruins and other archaeological features characteristic of the region, we may hope in time to gain not only an idea of prehistoric conditions but perhaps also an adequate explanation of the origin, the antiquity and the course of development leading up to a better understanding of the present status of aboriginal life in the region. (Nelson, 1913:63)

Nelson's supervisor, Clark Wissler (1915:395), indicated that the "plan was to take up the historical problem in the Southwest to determine if possible the relations between the prehistoric and historic peoples." Toward that end, the area chosen, apparently by Wissler, was what "seemed most likely to have been the chief center of Pueblo culture as we now know it" (Wissler, 1915:397).

Nelson's formal report of the first two seasons of excavation mentions only a few chronological observations:

> The surviving artifacts [recovered from the various sites sampled] were of the same types, with nevertheless a local and also a stratigraphic variation in general finish and decoration of the pottery.... [T]he execution of glazed ornamentation on pottery seems to have degenerated in late prehistoric times, but the artists continued to use the glaze of older days. (Nelson, 1914:111–112)

In a later popular report written while in the field, Nelson (1915) did not mention chronological issues, though an accompanying article by Wissler (1915:398) indicated that the

> net result of [Nelson's] work has been to make clear the chronological relations of the various ruins in the vicinity, which in turn enables us to determine their historic relation to the living peoples.... [T]he way is now clear to a chronological classification of many groups of ruins.

In his formal report on later excavations, Nelson (1916:162) stated that by the beginning of the 1914 season, he suspected he knew the chronological order of five types of pottery. Apparently, Nelson's initial efforts in 1914 to locate an undisturbed stratigraphic section so that he might study the chronology of ruins and pottery styles were thwarted (Nelson, 1916), but as we noted in Chapter 2, late in the 1914 field season he found the perfect section at Pueblo San Cristobal. Nelson had visited the stratigraphic excavations of Otto Obermaier and Henri Breuil in Spain in 1913, and had seen "levels marked off on the walls" of the excavations and participated in excavating Castillo Cave; this experience, according to Nelson, served as his "chief inspiration" for his excavations at Pueblo San Cristobal

(Nelson, in Woodbury, 1960b:98). His technique of excavating in arbitrary levels might have come from Europe, but certainly not the notion that superposed collections marked the passage of time. Everybody assumed that; they just thought that it had not been very much time, and hence was not worth worrying about, particularly when most technological and functional types of artifacts were found in both upper and lower strata.

Nelson excavated San Cristobal in arbitrary 1-foot-thick levels (Fig. 5.4) rather than in natural stratigraphic units, and he kept sherds from each level separate and identified and counted them by level. In his "Chronology of the Tano Ruins, New Mexico," Nelson (1916) presented the absolute abundances of each of five types, later termed "styles," of pottery from each of his ten levels, and in an accompanying table he adjusted observed sherd frequencies to account for different excavation volumes, both rather innovative procedures for the time, although he and Uhle had done the latter at Emeryville. His adjustment of observed frequencies, however, is unnecessary if the relative, or proportional, abundances of the artifacts are calculated, a standard practice in modern archaeology.

Kidder, who was working at Pecos Pueblo just to the east of where Nelson was working, also presumed that different excavation volumes compromised the usefulness of artifact frequency data, noting with regard to using strata as artifact collection units that the method "derogates from the absolute statistical value of the material, as the cuts, not being of exactly equal thickness, are not strictly comparable statistically" (Kidder and Kidder, 1917:340). It is unclear why Nelson, Kidder, and Uhle assumed that type frequencies must be adjusted to account for different excavation volumes. Maybe they were attempting to make the clearly relative time scale of superposition more absolute, as implied in Kroeber's (1919:259) assessment of Nelson's excavation: "Each foot of debris may be taken as representing an approximately equal duration of deposition, as indicated by the fairly steady number of sherds of all types found at each depth." Rates of pottery deposition would influence absolute abundances of sherds, but only if the rate of sediment deposition remained constant throughout the sequence. Clearly, not all of the relevant variables were being considered by Nelson, Kroeber, and Wissler.

It is the stratigraphic context of the pottery, not the potential conceptual or methodological implications of sherd frequencies for culture history, that is remembered by historians (e.g., Browman and Givens, 1996; Willey and Sabloff, 1993; Woodbury, 1960a,b). In a retrospective look at Nelson's work, his once-colleague at the American Museum Leslie Spier (1931:275) stated that Nelson's use of superposition was "the first exposition of a refined method for determining exactly the time sequence of archaeological materials in a primitive area," though in our view this is somewhat of an overstatement; Dall (1877), Uhle (1907), Harrington (1909b), and others had accomplished more or less the same thing (Lyman and O'Brien, 1999). Given that collecting artifacts from distinct vertical proveniences was not new, what was so novel about Nelson's approach to chronological questions and what was the source of that new approach?

Nelson (1916:162) stated explicitly that when he went to San Cristobal in 1914, he sought to test a suspected local sequence of four pottery types: "By the opening of the [1914] season, it was reasonably certain, both from internal evidence and from various general considerations, what was the chronological order of the four apparent pottery types, but tangible proof was still wanting." He knew or suspected the relative chronological positions of pottery types, but only by excavating at San Cristobal was he able to establish their relative chronological positions; all previous superpositional indications of chronology were "incomplete and fragmentary, each showing merely the time relations of two successive pottery types at some place or other in the total series of four or five types" (Nelson, 1916:163). We note that Nelson's use of the word "merely" demonstrates that he viewed the temporal implications of superposed pottery types as nothing out of the ordinary. Thus, Nelson never claimed that his stratigraphic excavation was particularly new, innovative, or revolutionary. More important, Nelson also did not claim that his analytical technique was new, innovative, or revolutionary.

Schenck (1926:170) believes that

> Both Uhle and Nelson proceed[ed] on the assumption that a lower level is older than a higher level no matter what its horizontal position.... Human activity should not be expected to produce a series of essentially homogeneous strata extending over the entire mound.

By our reading, even when he was working at Emeryville in 1906, Nelson was well aware of the problems with making precisely this assumption when he noted that the side of the shell mound that Uhle had worked on seemed to be older than the side that Nelson worked on, and more important he noted that "it seems clear that investigation of the Emeryville Shellmound cannot be considered satisfactorily complete until something like an open trench has been carried clear across the mound laying bare its core" (Nelson, 1906:18). After inspecting the stratigraphy of a Florida shell mound exposed by construction activities in 1917, Nelson (1918:86) was even more explicit about the differences between the depth and the stratigraphic position of artifacts: "Under certain conditions depth is an index of age [deeper being older].... But unless we take strict account of the order of deposition, the depth at which a given artifact occurs may signify little or nothing."

The important innovation found in Nelson's work is his demonstration that pottery types altered in absolute frequency through time in a pattern that he characterized as "very nearly normal frequency curves [that reflected the fact that] a style of pottery ... came slowly into vogue, attained a maximum and began a gradual decline" (Nelson, 1916:167). In Fig. 5.9 we present Nelson's data graphically as percentages of four pottery types by excavation level. Note that corrugated ware does not pass the historical significance test but that types I–III do. By using these three types, Nelson was able to measure culture change using not the then-typical qualitative differences in artifact assemblages such as the presence or absence of pottery, as had been done by Harrington and Wissler, or by plotting

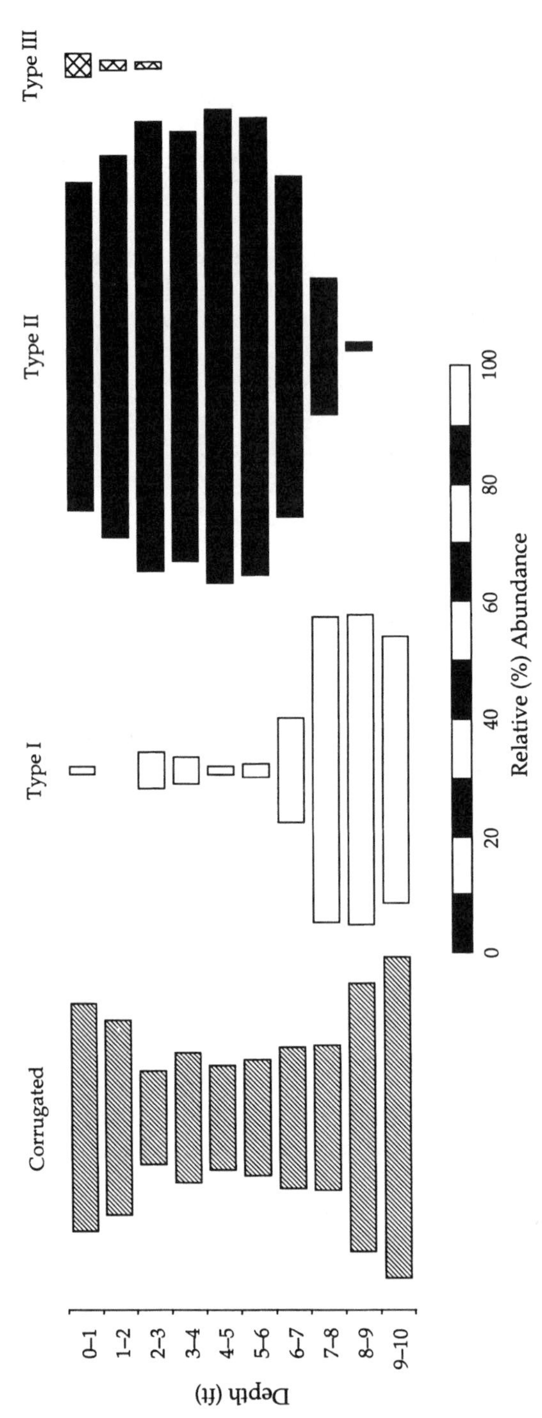

Figure 5.9. Nels C. Nelson's pottery data from Pueblo San Cristobal, New Mexico, showing the waxing and waning popularity of types. Note that Nelson believed, correctly, that the frequency of corrugated ware was not a good indicator of age. Note also the essentially monotonic frequency distribution of his types I–III (after Nelson, 1916).

frequencies of technological and functional types of artifacts against their vertical recovery proveniences such as he and Uhle had done with materials from sites in San Francisco Bay. Rather, Nelson detected culture change, and thus the passage of time by documenting, in revolutionary fashion, the changing frequencies of pottery types or styles. Wissler (1916b:195–196) suggested that such frequency changes in "specific styles in ceramic art [represented] stylistic pulsations." These represented, as Kroeber (1909:5) had noted a few years earlier, "passing change of fashion." The question thus becomes less one of wondering where Nelson got the idea to collect artifacts from vertically discrete units (many of his predecessors did that) but rather wondering where he got the idea to study the frequencies of nonfunctional and nontechnological artifact types. We believe it was a collaborative effort, and here the roles of Clark Wissler and particularly of Leslie Spier are critical.

Leslie Spier

In April 1912, the state legislature of New Jersey "authorized the commencement of archeological investigations under the direction of the Geological Survey of New Jersey. The Department of Anthropology of the American Museum of Natural History inaugurated [the] systematic archeological research in the summer of 1912" (Spier, 1913:676). Spier reported that by the end of the 1912 field season numerous prehistoric sites had been found, and he noted that those sites displayed a lack of homogeneity in their surface manifestations. Further, by the end of the 1913 field season, Spier (1913:679) observed that "the number of sites within [the] limited area [examined] is too large for all to have belonged to [the historical or colonial] period." That various evidence elsewhere (e.g., Harrington, 1909b) suggested time differences in archaeological remains constituted "proof that this [heterogeneity of cultural remains] is indeed a problem for serious study" (Spier, 1913:679). Just as Nelson had observed at about the same time, the bottom of the American shallow past seemed to be dropping away.

As Spier later noted, his efforts those first couple of years were worthwhile: "[A] fair amount of enterprise has succeeded in bringing to light several stratified sites" (Spier, 1918b:221); with such sites he could address the problem of heterogeneity in deposits. Spier began to examine the problem in 1913, when he excavated several sites in such a manner as to be able to note not only which depositional unit or stratum an artifact came from but also the vertical provenience, to the nearest inch, of each artifact. We suspect that he did so for several reasons. He was well aware of the earlier work of Henry C. Mercer (1897), who had dug a trench through a mound adjacent to the Delaware River and upon making some post facto stratigraphic observations concluded that the mound was stratified and contained the remains of "two village sites, set one upon the other—an upper and a lower" (Mercer, 1897:72). Mercer knew the superposed village sites were different in age, but he lamented that

> the upper site might have been inhabited one or five hundred years after the lower was overwhelmed. If, therefore, we sought for inference as to the relative age of the two sites,

> we could only hope to find it in a comparison of the relics discovered. Realizing this, the depth, position, and association of all the specimens found, and particularly their occurrence above or below the lines of stratification, was carefully noted. (Mercer, 1897:74)

Mercer (1897:82) did not provide details about how he excavated, but he did provide details regarding the different kinds of artifacts he found in the "two stages of occupancy. The layers prove a difference in time, short or long."

Mercer used superposition to argue that temporal differences accounted for the variation in artifact kinds comprising the stratum-specific assemblages. Spier (1913) immediately picked up on that observation, but he was working near the famous deposits around Trenton, New Jersey, that had played a role in the debate over the possible presence of glacial-age humans in North America (e.g., Volk, 1911). No doubt because of where he was working, Spier (1916b) was asked to review some of the published evidence that suggested there was a distinctive ancient culture in the general area. In his review, Spier argued that an apparent similarity of stratigraphic position might not be a valid indicator for "the identification of cultures" (Spier, 1916b:566). In particular, did the artifacts from the middle, yellow sand layer found in the Trenton area comprise a culture distinct from that of the historic peoples who had occupied the lower Delaware River and whose artifacts were said to be found in the black soil overlying the yellow sand? Had perhaps the artifacts in the yellow sand originally been deposited in the black soil, and as a result of some turbation process, ended up in the yellow sand? Here, Spier was echoing the earlier work of William Henry Holmes (1892, 1897), who had demonstrated the vertical movement of artifacts in eastern gravel beds, in the process showing that purportedly glacial-age tools were of more recent origin.

Continuing the American Museum's work in New Jersey, Spier excavated near Trenton in 1914; he excavated again in 1915, this time more extensively "as the stratigraphic relations dictated" (Spier, 1918b:169). He used the bread loaf excavation technique in 1914: "[T]renching proceeded by scraping the breast or face of the trench with a trowel. The depth below the plane of contact of [the uppermost] black and [middle] yellow soils ... and the lateral position of each [artifact] specimen was noted before it was removed" (Spier, 1918b:180). In 1915, Spier excavated several trenches using the bread loaf technique and excavated others in arbitrary 4-inch levels. Although he was not specifically asking chronological questions about superposed remains, what he was asking—had the artifacts in the yellow sand unit been deposited with that unit?—ultimately was critical for Americanist archaeology. Spier showed that the materials not only consistently occupied a particular vertical position within the yellow stratum but that those materials also consistently displayed a particular frequency distribution when plotted throughout that stratum vertically. He termed it a "typical frequency distribution" (Spier, 1916a:186; 1918b:185) that reflected a "normally variable series" (Spier, 1918b:192).

What is critical about this is the fact that Spier attempted to figure out why the artifacts displayed the vertically arranged frequency distribution that they did. The artifacts, he concluded, were not "intrusive from the [overlying] black soil"; had they been, "we would find the maximum frequency [of artifacts] at the plane of contact [stratigraphic boundary between the black soil and yellow sand units] with frequencies diminishing with depth" (Spier, 1918b:186–187). Instead, the maximum frequency was near the middle of the yellow sand unit. Wissler (1916b) and Spier (1918b) interpreted this frequency distribution as evidence of redeposition, an interpretation supported by a similar distribution of nonartifactual pebbles. Might this distribution "characterize the occupation of the site" (Spier, 1918b: 201)? Wissler and Spier showed that, when summed, Nelson's and Kidder's pottery from the Southwest did not display such a distribution. The unimodal frequency distributions exhibited by Nelson's sherds occurred only for "specific styles in ceramic art and not the typical distribution for the ceramic art of San Cristobal as a whole" (Wissler 1916b:195). That is, only sherds of particular "styles" (types) displayed a unimodal frequency distribution, not the total of all sherds irrespective of type.

Spier was even more clear on the significance of frequency distributions:

> In both pueblos [San Cristobal and Pecos] certain types of pottery give distributions of the normal type.... But these are not comparable to the Trenton series since they represent fluctuations in single cultural traits—stylistic pulsations [Kroeber's passing changes in fashion]—which attain their maxima at the expense of other similar traits. (Spier, 1918b:201)

Spier began plotting the absolute frequencies of artifacts late in 1913, about the same time that Nelson (1916) was first observing the relative stratigraphic positions of different kinds of pottery in the Southwest. Wissler wanted to show that the Trenton materials were naturally deposited in order to perpetuate the museum's long-standing position that the artifacts were not Pleistocene in age (Meltzer, 1983). Recall that Spier, Nelson, and Wissler were all working at the American Museum; the Trenton and San Cristobal analyses were temporally coincident; and absolute frequencies of artifacts were used in both analyses. In later analyses, Spier (1917a,b, 1918a, 1919) used percentages of artifact types, as did Nelson (1920). Nelson (1916:166) adjusted the absolute frequencies of his San Cristobal pottery to account for different excavation volumes. This brought his sherd frequencies in line with Spier's Trenton data and allowed Wissler and Spier to use that pottery data as evidence to argue that the Trenton materials had not been deposited by people.

A. V. Kidder

Kidder (1915:461) noted that a December 1914 (Anonymous, 1915) announcement indicated Nelson's excavation had revealed a deposit "so stratified that the relative ages of [four distinct types of pottery] could be ascertained," but he also lamented that "What the wares were has not yet been announced." Based on research he had carried out for his dissertation while a graduate student at Harvard, Kidder (1915:452) suspected what the sequence of pottery wares, or types, was, but "as yet no stratified finds have given us absolutely conclusive proof of this." His early suspicions regarding temporal sequence were founded on three things: (1) the association of certain pottery types with "nearly obliterated ruins of obviously greater age than any others in the region" (Kidder, 1915:452–453); (2) the geographic distribution of certain pottery styles; and (3) suspicions regarding the progression of pottery styles and technologies (Kidder, 1915:453–456). But he also noted the necessity of finding stratified sites in order to draw "reliable developmental or historical conclusions" (Kidder, 1915:461). Superposed collections would serve to confirm a local sequence rather than to create one. That was why he was so interested in the announcement concerning Nelson's work.

Kidder (1916:120) reported that Pecos Pueblo was chosen for study because historical documents indicated that it had been occupied from 1540 until 1840, and surface finds included "practically all types of prehistoric wares known to occur in the upper Rio Grande district." Occupation was thus believed to have been continuous and from "very early times." No other site then known in New Mexico and "available for excavation" seemed to have that attribute:

> [I] hoped that remains [at Pecos] would there be found so stratified as to make clear the development of the various Pueblo arts and to enable students to place in their proper chronological order numerous New Mexican ruins whose culture has long been known but whose relation to one another has been entirely problematical. This hope was strengthened by the fact that Mr. N. C. Nelson ... had recently discovered very important stratified remains at San Christobal a few miles to the west. (Kidder, 1916:120)

Similar deposits at Pecos would allow comparative analyses and the extension of Nelson's chronology, which ended at 1680, when San Cristobal was abandoned, into the middle of the nineteenth century.

Early in his excavations, Kidder (1916:122) recognized that pottery types in the lower levels of his trenches were "markedly different from [those] at the top and that there were several distinct types between." The stratigraphic positions of the types in one part of the site are shown in Fig. 5.10. Not all excavations at Pecos Pueblo were undertaken with close attention to superpositional relations. Rather, such relations were observed by

> tests made at different points as the [excavation] advanced. The tests consisted of the collection of all the sherds in a given column of débris, the fragments from each layer being placed in a separate paper bag bearing the numbers of the test and of the layer. (Kidder, 1916:122)

During the first year's field season, the summer of 1915, those tests employed arbitrary levels that were 1- to 1.5-foot thick, but when it was apparent that the arbitrary levels split strata, "a new bag was started," which apparently meant that Kidder paid attention to the stratigraphic provenience of the artifacts found within

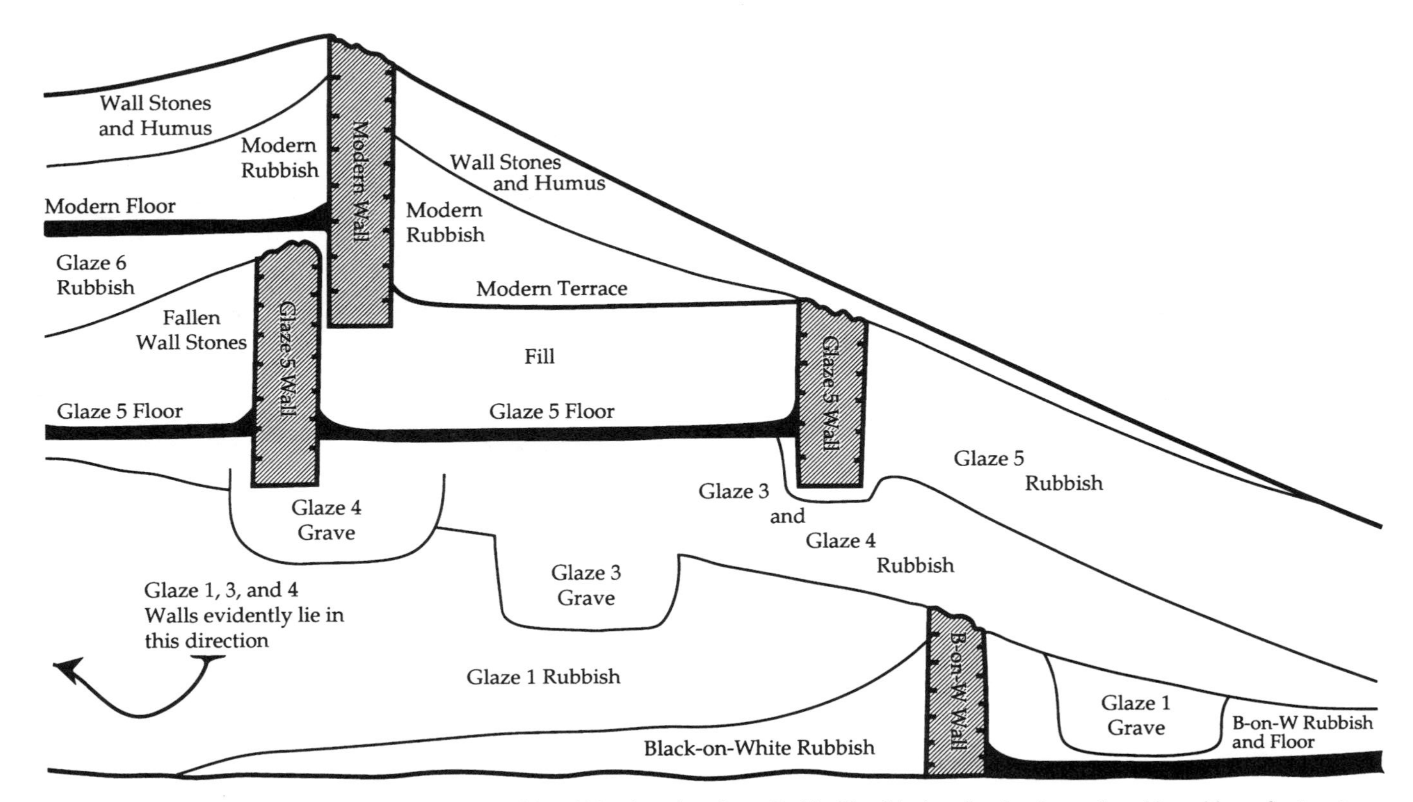

Figure 5.10. A. V. Kidder's cross section through one of the midden deposits at Pecos Pueblo, New Mexico, showing the stratigraphic positions of pottery types and architectural features (after Kidder, 1924).

his arbitrary levels. He was careful to place his tests only in areas that appeared to be undisturbed, abandoning tests "when it became clear that a grave shaft had disturbed the original deposition of the refuse" (Kidder, 1916:122). By the second field season (1916), Kidder was excavating in natural stratigraphic levels (Kidder and Kidder, 1917:340).

In a detailed discussion of his excavation methods written after the completion of his fieldwork, Kidder indicated that when strata were particularly evident in vertical exposures at Pecos, four procedures were followed: (1) "a column, its size determined by the extent of the deposit, was isolated by exposing two, three, or even all four of its sides by trenching from surface to bottom. A horizontal base line was established toward the center of this column, for measuring purposes"; (2) the vertical faces of the columns "were then carefully scraped to a vertical face to reveal every line and band in the column, and the limits of each cut to be made were fixed along such lines by placing pegs at frequent intervals to guide the workmen, who were instructed to cut each layer precisely down to the row of pegs"; (3) a stratigraphic profile was then drawn to show the "position and thickness of each layer. In dividing the column into layers care was taken to follow a natural division ... rather than an arbitrary line. The resultant layers were not always of equal volume ... but they did represent the actual structure of the column"; and (4) excavation of the column proceeded "layer by layer, working from the surface downward.... All objects found in a layer were of course kept apart [and] labeled with the cut number they represented, for later study.... The strata alone determined the number of cuts made" (Kidder, 1931:9–10).

Kidder, unlike his predecessors Nelson and Gamio, both of whom used arbitrary levels, excavated portions of Pecos after the first field season in natural stratigraphic layers. Kidder's biographers (Givens, 1992:50; Willey, 1988:307; Woodbury, 1973:43) and historians of archaeology (e.g., Willey and Sabloff, 1993:103–105) have suggested or implied that Nelson influenced Kidder relative to paying attention to superpositional relations of artifacts. However, it probably is more accurate to suggest that it was Kidder who perfected the technique and modified it to focus on natural stratigraphic units. Although Kidder is remembered for peeling back individual strata and collecting artifacts from each natural unit, he made two other seldom remarked but nonetheless important contributions to the measurement of ordinal scale time.

Recall that Nelson (1916:167) described frequency distributions as "very nearly normal frequency curves" and that such distributions were to "be expected." Kidder mimicked Nelson's analytical technique but modified it in the process. Kidder not only listed the absolute frequencies of pottery types against their vertical provenience in tabular form—the mimicking part—but he also graphed the changes in relative frequencies of his pottery types against his excavation levels (Kidder and Kidder, 1917)—the innovative part. This analytical technique, later referred to variously as "ceramic stratigraphy" (Willey, 1938,

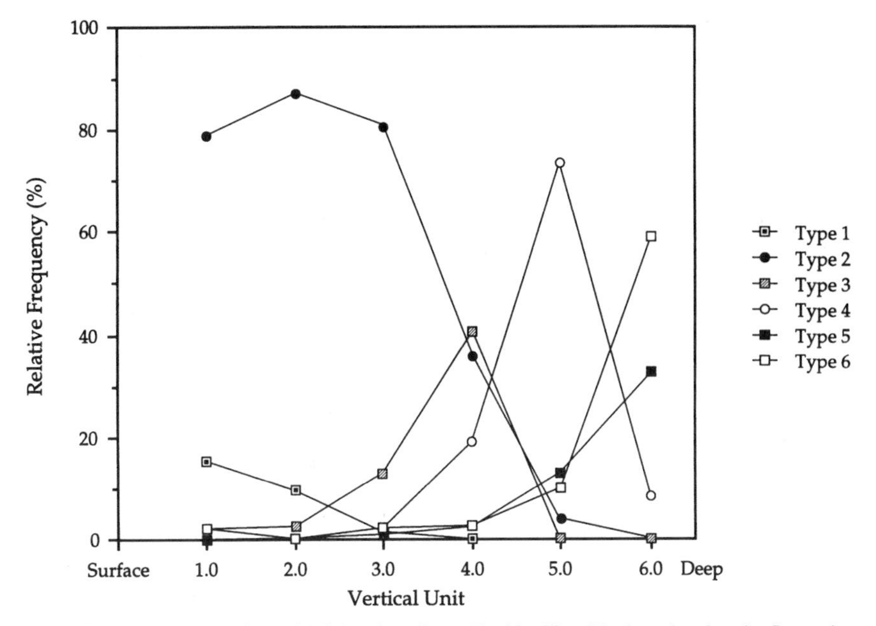

Figure 5.11. A broken stick graph of data from Pecos Pueblo, New Mexico, showing the fluctuating frequencies of types over vertical space (time) (after Kidder and Kidder, 1917).

1939) or "percentage stratigraphy" (Ford, 1962; Willey, 1939:142), was his first contribution. His second was to graph the changes in frequencies of types rather than to merely present the data as a table of numbers. We have redrawn one of the graphs in Fig. 5.11. Upon inspection of the graphs, Kidder and Kidder (1917:341) noted that many, but not all, types displayed "approximately normal frequency curves." They echoed Nelson and Wissler and interpreted such curves as "indicating that each [type] had a natural rise, vogue, and decline" (Kidder and Kidder, 1917:349): the popularity principle.

What *Was* the Revolution?

From the preceding it should be clear that the so-called "stratigraphic revolution" did not reside in a change in excavation technique. Rather, it resided in a shift in how variation in artifacts was measured; this shift in how artifacts were classified (and interpreted) in the second decade of the twentieth century made culture change and thus time visible. As Nelson (1919c:134) observed a few years after what we might most accurately call the "historical-type revolution," the research of Kidder, Kroeber, Spier, and Nelson himself indicated that pottery was "the most readily available single trait through which to trace the preliminary outlines of Pueblo history." Rather astutely, Nelson (1919c:134) noted that "ceramics, or pottery," worked well for measuring time "mainly because of its variability." He elaborated that "utilitarian objects," or what we would term functional classes, "would not serve because they are much the same the world over," whereas with respect "to both form and decoration, [pottery] gives infinite opportunity for the expression of individual taste and talent and the resulting differentiation in the product from place to place and from time to time is the key to the [chronological] problem." Finally, once a sequence of pottery styles had been worked out, then "associated architectural developments and other allied traits will come very near to falling into their own proper places" in time.

Two years later, Wissler (1921:15) remarked that pottery was "so plastic in origin as to give almost free rein to styles of artistic expression and the use of decorative technique.... The [chronological] record, therefore, is in the simple potsherds, and our problem was to find a method for reading it." He went on to state that "the best indexes to time differences are the changing styles of pottery" and that "pottery characters are the most accessible and lend themselves most readily to the method of superposition" (Wissler, 1921:23). But the perception of what happened in the second decade of the twentieth century had already begun to shift. Wissler (1917b:100) had earlier stated that when asking questions of chronology, an archaeologist "must actually dissect section after section of our old Mother Earth for the empirical data upon which to base [our] answers." He failed to note the role of typology. A few years later, Kidder (1924:45–46) indicated that the "ideal form of chronological evidence is provided by stratigraphy, i.e., when remains of one type are found lying below those of another." He devoted three paragraphs to the importance of stratigraphy, but only two sentences explicitly referred to the role of typology, and those—as the one quoted here—failed to underscore its critical significance.

After the Revolution: Measuring Time with Strata

No doubt as a result of statements such as Wissler's and Kidder's, the primary analytical unit of an archaeologist interested in time became vertically bounded assemblages of artifacts variously referred to as components, cultures, or occupations (e.g., Colton, 1939; Gladwin and Gladwin, 1934; McKern, 1939; Rouse, 1955). Stratigraphic excavation was now not just good field technique but something that was of great importance for measuring time. Superposition was perceived, as it had been from the beginning, to be a more trustworthy indicator of time than was seriation (e.g., Rowe, 1961). Eventually the use of vertical units, whether of the depositional or arbitrary sort, to measure time resulted in reinforcement of the perception that artifacts within vertically bounded recovery units comprised real, discrete entities rather than accidental (results of deposition) chunks of a cultural continuum (Lyman and O'Brien, 1999). Strata no longer were geological depositional units; they were archaeological and cultural units, and thus could be "reversed" (e.g., Hawley, 1937) or "mixed" (e.g., Phillips *et al.*, 1951), the latter no doubt the result of Holmes's (e.g., 1892, 1897) earlier observations that supposedly late Holocene artifacts could occur in late Pleistocene deposits.

The difficulties of stratigraphic reversal and mixing were perceived because archaeologists strayed from the geological notion of superposition. Remember, the principle of superposition relates only to the chronological order of strata deposition, not to the age of the sediments or particles (including fossils and artifacts) making up the depositional layer. The order of deposition might or might not be related to the age of the sediments. Failure to keep the two straight resulted in all sorts of confusion, such as is evidenced in the following statement by John McGregor:

> to the most uninitiated it is obvious that the lowest layer must have been laid down first to support those resting upon it. Thus the lowest layer is oldest, the next above younger, and so on to the top, where the most recently laid down layer, or stratum, is found. A study of the contents of these layers will then give some idea of the relative ages of the strata.... The most recent or latest material will come from the top layers; the earliest or most ancient, from the bottom. The archaeologist then need only arrange these objects in the order in which they were uncovered from the bottom up and he has a sequential evolution of the culture of the people who made them. (McGregor, 1941:45–46)

McGregor had the order of deposition part correct, but he confused the age of sediments with order of deposition. We also can see in McGregor's comments the notion that layers can be interpreted as cultural units, each reflecting a different culture. This is a critical point that warrants elaboration.

Stratigraphic examination of excavated strata has long been a mainstay of archaeological efforts at chronological ordering and will always remain so, because superposed artifacts offer a ready means of measuring the passage of time. However, given the problematic nature of the relation between strata and artifacts in the strata, caution clearly is called for in making chronological inferences. When this issue is ignored, there is every reason to be less than enthusiastic about a purported chronology built on stratigraphic evidence. With respect to Americanist archaeology, once it was clear that culture change was analytically visible in the archaeological record, stratigraphic excavation became commonplace, and excavation techniques subsequently were described more explicitly in the literature than they had been previously (see Praetzellis, 1993, for a history). Superposition and stratigraphic excavation provided empirically sufficient standards by which one could test a suspected sequence of artifact forms. The necessary standard that superposed artifacts represented the passage of time was dealt with by referring to selected strata—depositional units—as mixed or reversed in chronological order when two or more sequences of artifacts failed to align with one another. Such ad hoc rationalization attends the fact that superposition cannot order the contents of strata without a consistent relationship between the age of the contents and the order of deposition and underscores the commonsensical approach to chronology that in many ways still typifies archaeology.

It probably was only natural that strata would begin to be equated with cultures. In fact, despite the innovative work relative to chronological ordering that took place in the teens, seeds of the strata as cultures view already were beginning to sprout. Remember how Manuel Gamio had superimposed the cultures of the Valley of Mexico on his cross section at Atzcapatzalco (Fig. 5.8). In similar fashion, Wissler (1916b:190) cited "Mr. Nelson's decisive chronological determinations in the Galisteo Pueblo group" as an excellent example of how a scientific result "depends upon the method of handling data." Most importantly, from the perspective of stratigraphic excavation, in the first edition of his classic *The American Indian*, Wissler (1917a:275) remarked that the

> uninterrupted occupation of an area would not result in good examples of stratification [of cultures], but would give us deposits in which culture changes could be detected only in the qualities and frequencies of the most typical artifacts; for example, Nelson's pottery series from New Mexico.

Here was a hint that strata, as units of time and as units of associated cultural materials, would come to have much more significance to the culture history paradigm than perhaps was originally intended. In short, the presence of strata, that is, discrete and distinct depositional units, implied a discontinuous occupation by multiple successive cultures. As Gerard Fowke (1922:37) noted several years later, the "intermittent character of occupancy is ... shown by the distinct segregation of numerous successive layers of kitchen refuse."

Most archaeologists in the 1920s (e.g., Gamio, 1924; Harrington, 1924; Hawkes and Linton, 1917; Judd, 1929) spoke of the remains of different "cultures" being found in distinct strata, with each culture being represented by the artifacts associated within a vertical unit, regardless of whether that unit was arbitrarily defined by metric levels or by the boundaries of natural depositional units. This is not surprising given that each such unit, particularly natural depositional units, were thought to represent an "occupation," whatever that was. Wissler's and Fowke's suggestions were reinforced by the notion that a small site represented a brief occupation by one culture, and thus was to be preferred when one sought to identify the constituents of a cultural phase. Ultimately, the view that each culture was preceded and succeeded by other such units was reinforced by the use of vertically superposed depositional units as units from which artifacts were collected. Such a collection strategy would characterize archaeology beyond the 1960s when the culture-history paradigm under which the measurement of time had been so important was replaced by processual archaeology.

MEASURING TIME AT GATECLIFF SHELTER, NEVADA

One project that grew out of the processual archaeology heyday was David Hurst Thomas's work in Gatecliff Shelter, located in the Monitor Valley of central Nevada. We focus briefly on that project because it is an excellent example of how an archaeologist goes about interpreting a stratigraphic sequence. The 10-plus meters of deposit in Gatecliff (Fig. 5.12) were excavated over seven field seasons between 1970 and 1983. Numerous stratified sites in the Great Basin had been excavated over the years, but none in the region in which Thomas was working. Based on those previous excavations he had a fairly good idea of some of the basic changes in projectile point shape, changes that could be used to mark certain archaeological periods; but there was no reason to suspect that those changes had occurred everywhere in the Basin simultaneously. Gatecliff Shelter seemed to hold the promise of containing stratified deposits that if excavated carefully might yield the chronological information Thomas needed. There were, of course, numerous other sources of information that potentially were buried in the deposits, prehistoric foodstuffs, for example, but the big draw was the shelter's chronological potential, and Thomas's excavation paid off handsomely.

We use Thomas's discussion of Gatecliff as an example of how one goes about making stratigraphic interpretations, and we do so not only because of the spectacular nature of the deposits, well documented in the photographs and line drawings in his final report (Thomas, 1983), but also because of the careful manner in which Thomas made his interpretations. The monograph is a model of how to report an excavation. Moreover, it is a textbook in terms of how one deals conceptually with archaeological materials embedded in geological sediments. Fifty-six strata were identified in the wall profiles at Gatecliff. Notice how Thomas defined a stratum:

> A stratum is a layer of more or less homogeneous or gradational sedimentary material, a lithological unit visually separable from adjacent layers by a discrete change in the character of the material deposited, by a sharp break in deposition, or both. A break between adjacent strata may be marked by surfaces of erosion, non-deposition, an indication of pedogenesis or other abrupt changes in character. A stratum may, however, consist of multiple beds; a bed must be at least [one centimeter] thick. (Thomas, 1983:46; adapted from Gary *et al.*, 1974:698)

Further, Thomas continuously emphasized the distinction between stratum and horizon:

> Throughout this volume, *stratum* is used to imply a physical geological context, whereas the term *horizon* denotes a cultural unit.... The 56 geological strata are intended to provide the basic units of stratigraphic description for Gatecliff Shelter.... Let us emphasize once again that these *strata* are units of physical geology, whereas *horizons* are units defined on the basis of cultural context. (Thomas, 1983:46)

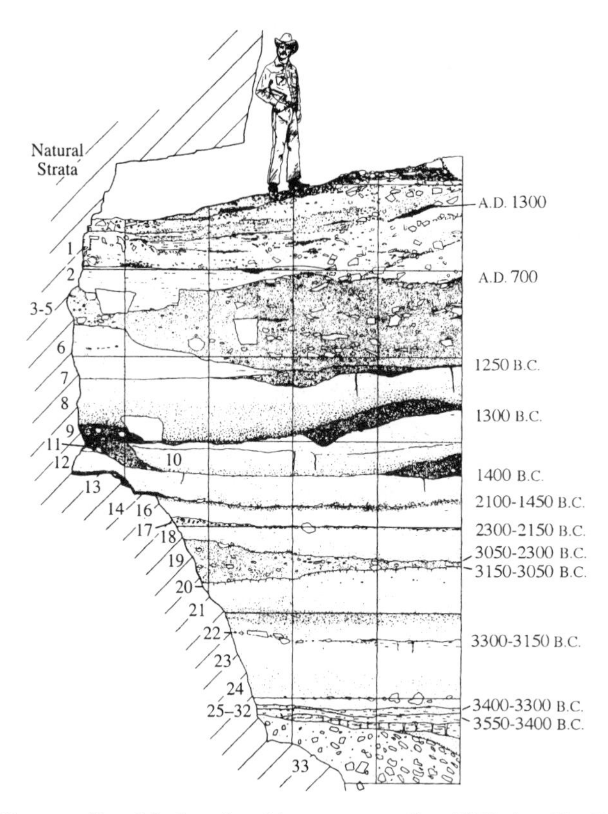

Figure 5.12. Master profiles of final stratigraphic exposures at Gatecliff Shelter, Nye County, Nevada (looking east), showing relation of 33 of 56 stratigraphic units (above) to 16 cultural horizons (next page). Each grid unit is one meter square (after Thomas, 1983) (courtesy American Museum of Natural History).

The depositional history of Gatecliff Shelter was difficult to unravel and took considerable time and patience on the part of the stratigraphers. The uppermost 33 strata are illustrated in Fig. 5.12, but this at best is a simplified version of a complex situation, as Fig. 5.13 makes clear. Deep sites located on flood plains are often easier to interpret than rock shelters and caves because of the nature of the deposition. For example, a stream or river overflows its banks and spreads sediments fairly evenly across the landscape. As a rule, coarser sediment particles are deposited closer to the stream and finer particles farther away, but overall the distribution of sediments is fairly even, creating horizontal beds across the landscape. In a confined space, however, wind- and waterborne sediments tend to get piled against walls, and they are always subject to churning by humans as well as

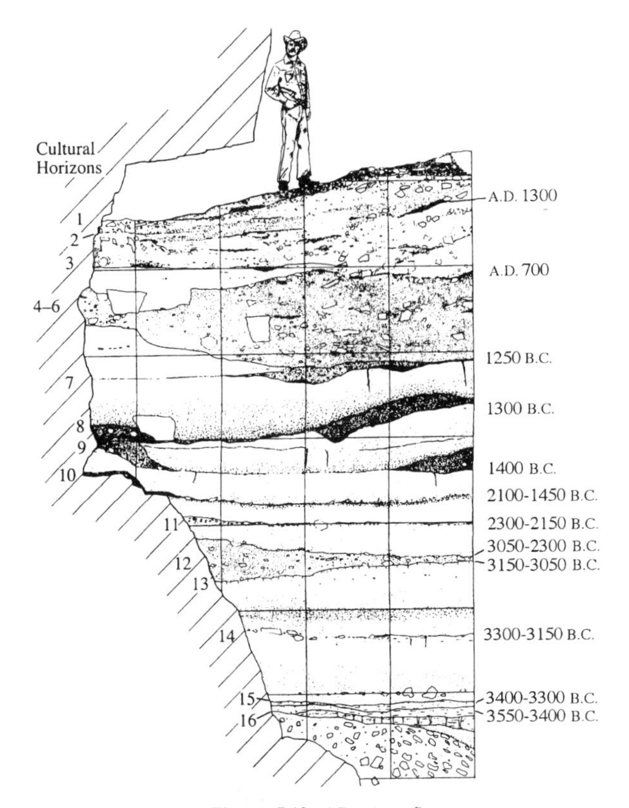

Figure 5.12. (*Continued*)

by burrowing animals looking for a warm (or cool in the summer), dry place to escape the elements.

Although people live in caves and rock shelters, many of their activities are carried out in front of them. Over time, a talus slope builds up in front of the opening, extending downslope from the edge of the overhang. These areas are usually rich in artifacts and attract the attention of archaeologists. But talus slopes are not nice, flat surfaces, and instead dip downslope. Since a talus slope grows in conjunction with what goes on inside a shelter and on the overhanging roof, fairly horizontal surfaces under the overhang suddenly plunge when they reach the outside. This is evident in Fig. 5.14, which shows horizontal strata superposed over dipping strata. Another feature that showed up in the Gatecliff talus profiles was

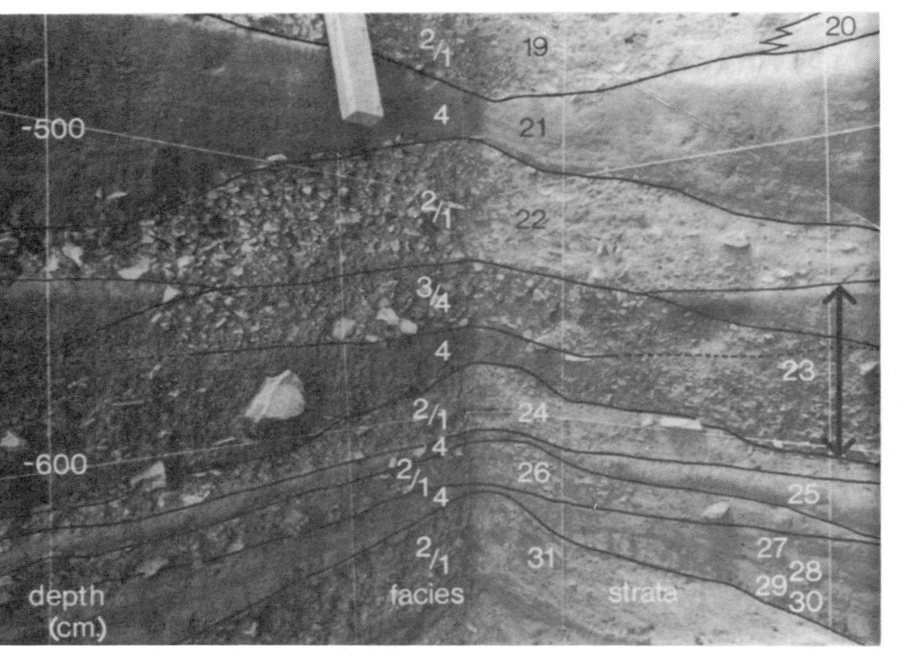

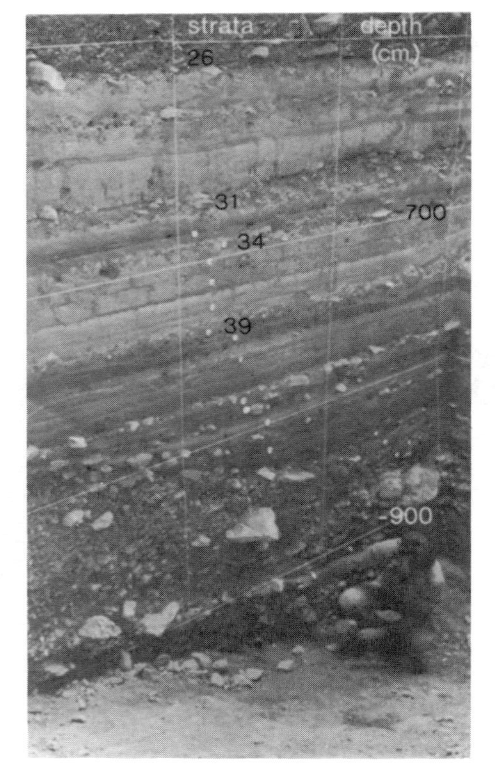

Figure 5.13. Photographs of strata at Gatecliff Shelter, Nye County, Nevada. The photograph at the left illustrates the correlation of facies—deposits that pinch out horizontally—and stratigraphic units (strata). In the photograph at the right, David Hurst Thomas points to stratigraphic unit 55, which contained a band of Mazama tephra. Independent dating at other localities has demonstrated that the tephra dates to around 4900 BC and the eruption of Mount Mazama in Oregon. Thus the tephra provides an anchor point for the geoarchaeological sequence at Gatecliff Shelter (from Thomas, 1983) (photo courtesy American Museum of Natural History).

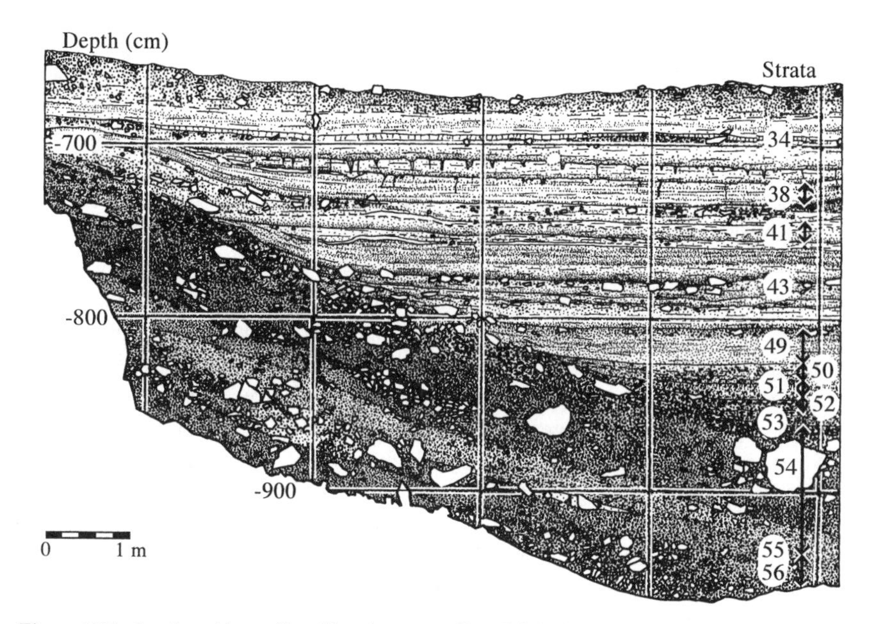

Figure 5.14. Stratigraphic profile of basal strata at Gatecliff Shelter, Nye County, Nevada (looking south). Notice the steep slope of the basal deposits and the horizontal nature of the overlying deposits (after Thomas, 1983) (courtesy American Museum of Natural History).

deposits of coarse sediments—sand, angular rocks, and the like—that had slid downslope from above the shelter during heavy rains.

Figure 5.12 also shows the vertical positions of 16 cultural horizons, or archaeological units, identified by the stratigraphers. They ranged in magnitude from a stratigraphically distinct flake scatter to an artifact-rich rubble unit up to 50-cm thick. Distinct cultural units, or horizons, were sometimes recognizable within a single geological stratum, for example, in strata 1 and 5 (Fig. 5.12), whereas in other cases multiple geological events were recognizable within a single cultural unit, for example, in strata 6 and 7. Thomas equated the terms horizon and occupation, which is problematic unless it can be demonstrated that all artifacts within a horizon were left by a group of people who never gave up occupancy of the shelter during the period when those artifacts were being incorporated into the archaeological record. In some cases this seems fairly evident (Thomas, 1983), but in other cases it is questionable. That point aside, Thomas did a remarkable job of documenting how "the bands of cultural debris interleaved with the purely geological strata" (Thomas, 1983:172).

Thomas lumped his 16 cultural units into a series of distinct components, following the standard definition supplied by Gordon Willey and Philip Phillips (1958:21): a local manifestation of a cultural phase, which is a time–space unit used to segment a regional archaeological record. Thomas (1971) had previously defined the phases during his settlement pattern survey of the Reese River valley. Note the way in which the Gatecliff components were integrated into the phase system:

> The Gatecliff components correspond in some cases to a single cultural horizon (e.g., Horizon 1 represents the entire Yankee Blade component at Gatecliff). But in several cases the stratigraphic separation is sufficient so that a given component may be subdivided into several horizons; the Reveille phase, for instance, is represented by Horizons 4, 5, 6, 7, and part of 8. (Thomas 1983:172)

One result of the Gatecliff excavation was the recovery of more than 400 projectile points that fell into types established previously for the Great Basin. Remember that one of Thomas's main objectives in excavating the shelter was to check the applicability of previous chronologies for the Great Basin that had been built primarily around changes in the hafting region of projectile points. Figure 5.15 illustrates the vertical ordering of projectile point frequencies by type. Given their orderly arrangement, we would say that most of them pass Alex Krieger's (1944) "historical significance test"; that is, each type came into style, reached a point of maximum popularity, then disappeared. The only type that appears to fail the test is the Humboldt Concave Base type, which seems to have two modes of popularity: one in horizon 12 and another in horizon 4. The dual-mode phenomenon might be tied to sampling error, or this type may be a poor historical type. Whatever the case, a relatively clean chronology of projectile point forms resulted from careful stratigraphic excavation.

THE FINAL PROOF IS IN THE SPADE, BUT ...

As we have noted and the heading of this section suggests, using superposition to measure time is an important part of an archaeologist's tool kit. But the adoption of stratigraphic excavation as the way to collect artifacts results in time being measured discontinuously, insofar as time is rendered as differences between superposed collections of artifacts. There are a number of reasons for this, but we will focus on the two most important one. First, geological processes "operate over long time spans" and, in general, are relatively slow and gradual (Dean, 1993:59). Cultural processes operate relatively much more rapidly, and thus the result typically is that geological units such as strata "span longer periods than do units of archaeological analysis" such as phases, occupations, tool kits, and the like (Dean, 1993:60). Although such can sometimes be controlled analytically, that is not always possible. Second, if artifact types are extensionally derived from collections in hand and the decision regarding which specimens to include in a collection is based on stratigraphic boundaries, then depositional history rather than cultural history is the determining factor in what constitutes a type.

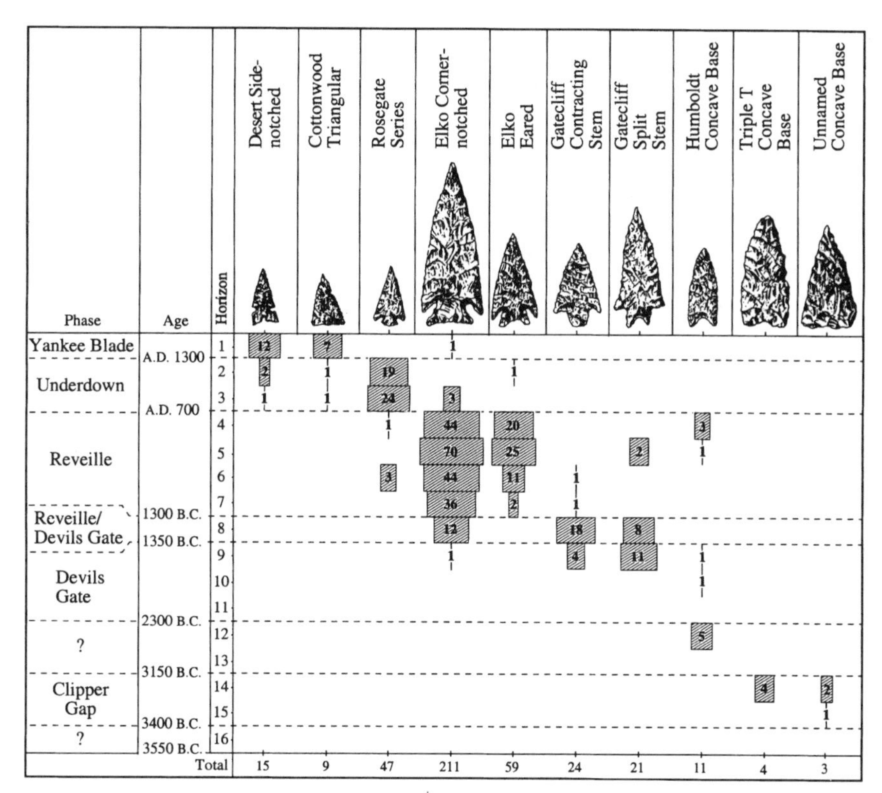

Figure 5.15. Stratigraphic distribution of projectile points from Gatecliff Shelter, Nye County, Nevada, that could be placed into previously existing types. Numbers in boxes are frequencies of points by type and level (after Thomas, 1983) (courtesy American Museum of Natural History).

Of all the archaeologists whose work we are familiar with, only James A. Ford has noted the second problem, and he was rather clear in his assessment of it:

> We must insist that the vertical separation of potsherds and other cultural materials only by the observable breaks in the deposit would be an archaeological variety of cataclysmic geology. If the observable lines of [stratigraphic] demarcation are significant events in the history of the site, then the collections from the several strata will give us the information to distinguish [cultural] periods. We can say that the lowest deposit is period A, followed in turn by B, C, etc. By this procedure, we have allowed the history to be separated into periods by chance historical events.... The chance that a neighboring site, occupied for the same span of time, was subjected to the same sequence of events seems remote.
>
> However, we are by no means justified in assuming that every observable line of demarcation represents an important event; it may or may not.... The only way to determine which condition prevailed is to excavate in the thinnest arbitrary levels

possible. The stratification must be recorded carefully. After the pottery from these arbitrary levels has been graphed, it can be correlated with the visible stratification, and the significance of the stratification will become apparent.

All archaeological techniques must be used reasonably and logically. It is absurd to dig automatically in ten-centimeter levels as it is to separate collections only by visible stratification. (Ford, 1962:45)

The procedure of correlating evidence of cultural change with stratigraphic boundaries is described by Phillips, Ford, and Griffin (1951), and one of their examples is shown in Fig. 5.16. The debate among those authors regarding whether to believe in the results of Ford's analysis (the battleship-shaped frequency curves

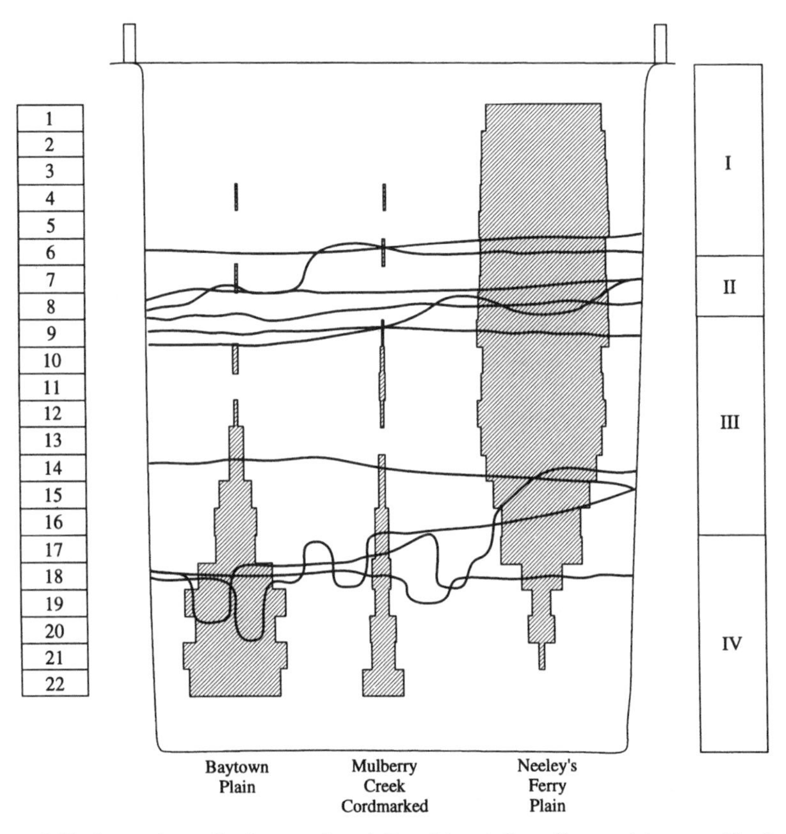

Figure 5.16. Composite profile diagram of cut A, Rose Mound, Cross County, Arkansas, with relative abundances of three pottery types by excavation level superimposed. The column of numbers in boxes on the left refer to 10-cm excavation levels. Roman numerals on the right denote idealized strata obtained by averaging the four walls of the 2-m^2 excavation unit. Heavy undulating lines in the center are composites of strata boundaries using all four profiles (after Phillips *et al.*, 1951).

in Fig. 5.16) or the results suggested by clearly superposed materials only (omit all arbitrary levels in Fig. 5.16 that fall near stratigraphic boundaries and thus are potentially "mixed") is significant in the present context. It underscores that if one views change as continuous and gradual, such as Ford did, then superposed materials are supplementary rather than primary to the measurement of the rate of culture change; the role of superposition is to indicate the direction of the flight of time's arrow and to confirm that the types constructed are measuring time. But if one views strata as primary not only to the detection of time but also to the measurement of the rate of change, then change is discontinuous and its rate punctuated. Using sets of artifacts to identify cultures, phases, and the like, when "setness" is based on stratigraphic boundaries, rests on a typically unspoken assumption:

> The usual, but seldom stated, assumption of ceramic phases is that ceramics go through periods of relative stability alternating with short periods of rapid change. The idea is that an overall pattern of style characterizes the ceramics of a given phase. This pattern changes relatively little during the phase compared to the more dramatic changes which occur fairly rapidly at the end of the phase and the beginning of the next, resulting in a different stylistic pattern which characterizes the ceramics of the next phase. (Drennan, 1976:298)

In short, time is measured discontinuously if cultural phases are constructed from a set of components, which in turn were identified on the basis of stratigraphic boundaries.

That superposed collections of artifacts often measure time is granted by all archaeologists. And the fact that such are often discontinuous measures has been recognized for decades. In an early statement to this effect, Meighan (1959:203), for example, noted that "seriation methods provide a more detailed picture of temporal change than stratigraphy and field observations alone." Fifteen years later, Fred Plog (1974:45) noted that change is forced to reside only at the boundaries between cultural phases, stages, and similar units when such are the primary analytical units and these units in turn have stratigraphic boundaries. We do not seek to discourage the use of stratigraphic excavation, and in fact that is why we presented Thomas's work at Gatecliff Shelter as an example of good stratigraphic work. Rather, we simply want to emphasize that such a procedure, if not tempered by qualifications such as those offered by Ford, will produce a much more discontinuous measure of time than may be desirable. As will become clear in the next chapter, superposition is not the only tool for relative dating that can produce discontinuous measures of time.

6 Cross Dating

The Use of Index Fossils

Geologists long have used fossils included in units of deposition to correlate those layers across space, often vast amounts of space. Familiar names in stratigraphic circles—William Smith, Adam Sedgwick, and Sir Roderick Impey Murchison in England, Georges Cuvier and Alexandre Brongniart in France—are credited with introducing the notion of such correlation into the broad field of geology during the first half of the nineteenth century (Rudwick, 1996). Since that time, the subfield of biostratigraphy has blossomed into a mature discipline in its own right, complete with competing strategies and philosophical arguments. Several things are evident in the biostratigraphic literature. First, biostratigraphers keep the distinction clear between the relative age of strata and the age of things in the strata, pointing out that at a single collection site, all that can be documented with certainty is succession, that is, which fossils are higher or lower in strata relative to other fossils. As C. W. Harper (1980:240) notes, "we cannot even say with certainty that a fossil specimen represents the same age as the narrow stratigraphic horizon which contains it."

Second, biostratigraphers use only organisms likely to be contemporaneous with the depositional event represented by a stratum when they attempt to correlate strata in time on the basis of their fossil content (see Hancock, 1977, and Mallory, 1970, for historical reviews). How is such likelihood determined? Their choice of index fossils is empirical; one has to examine many exposures to determine which kinds of organisms were unique to and diagnostic of a particular stratum and time period (Hancock, 1977). Third, biostratigraphers are well aware of the time-transgressive nature of the fossil record. That is, simply because a taxon had a high potential for geographic dispersal does not imply that it appeared everywhere at once, nor does it imply that it disappeared everywhere at once. Fourth, in arguing that similar successions of taxa occurred in separate localities, which must be so to allow temporal correlation of strata in different localities, biostratigraphers view the phylogeny of the taxa as irrelevant. This is the case because "evolutionary relationships of taxa are not known but can only be inferred, usually with less credibility than we can infer their relative ages" (Harper, 1980:242).

Recalling our discussion of heritable continuity in Chapter 3, the last may seem strange, but it is not, for as Niles Eldredge and Stephen Jay Gould (1977:25–26) indicate,

> *All* the many kinds of biostratigraphic units ever proposed share the simple assumption that similar organisms in different outcrops imply some kind of equivalence of their enclosing matrices. Even if we take this equivalency to imply time, is it really necessary for us to realize that evolution underlies our attempts to subdivide geologic time and accounts for the vertical, and even the horizontal, changes in the fossil content of strata? A thorough grasp of evolutionary theory has not been essential to the working biostratigrapher.... Some of the most intriguing, durable, and successful work in biostratigraphy has been performed by paleontologists holding rather dubious views on the nature of the evolutionary process.

Likewise, archaeologists can perform successful biostratigraphic work without being evolutionists. Why do we call what archaeologists do biostratigraphic? Because they, like paleontologists, study artifacts, the hard parts of past phenotypes (Dunnell, 1989; Leonard and Jones, 1987; O'Brien and Holland, 1990, 1995; O'Brien *et al.*, 1998). Although we freely admit that good biostratigraphic work can be done in archaeology without being grounded in evolutionism, we strongly agree with Eldredge and Gould (1977:26) that "a more accurate and complete picture of evolutionary mechanisms would benefit biostratigraphy by sharpening our practices and helping us to weed out techniques based on idealized, if not downright spurious, notions of evolution." We thus refer to evolutionary theory throughout this chapter to clarify the assumptions underlying archaeological cross dating and index fossils.

Biostratigraphers use pattern repetition in the sequences of distinctive morphologies of fossils found in numerous localities as checks on the accuracy of a proposed correlation. They are not concerned when told such things as

> It cannot be denied that from a strictly philosophical standpoint geologists are here (in using fossils to determine relative ages of strata) arguing in a circle. The succession of organisms has been determined by a study of their remains buried in the rocks, and the relative ages of the rocks are determined by the remains of the organisms they contain. (Rastall, 1956:168)

They are unconcerned because what can be circular about searching for recurrent, nonrandom patterns across a series of localities? If such patterns are identified, the biostratigrapher uses them to infer the relative ages both of the fossils and of the local strata that contain them (Harper, 1980:246). There is nothing circular about this; it is the way science works.

Archaeologists typically have omitted the stratigraphic cautions and caveats offered by biostratigraphers for two reasons. First, the rationale linking organisms to depositional units is implicit (Hancock, 1977; Mallory, 1970, and references therein), and thus cannot be readily converted to an application linking artifact manufacturers to depositional units. How does one, for example, identify an ethnographiclike occupation in the archaeological record (e.g., Dewar, 1992)? As we noted in Chapter 5, a stratum containing artifacts is not necessarily the same as an occupation, although this seems to be at least part of the source of the second reason archaeologists ignore the caveats of biostratigraphers. Archaeologists focus on artifacts, so it is not surprising that they label some stratigraphic sequences "reversed." The depositional sequence is not reversed, but this point became irrelevant in Americanist archaeology as soon as strata became both units of time and units of association, usually referred to as cultures, or occupations. To identify and correlate such units, one needed index fossils, and certain artifact types quickly filled the bill. A more precise term for archaeological materials would be index artifact, but following past and present use we go along with index fossil.

A. V. Kidder's comments in the introduction to his final volume on the pottery from Pecos Pueblo tend to epitomize much of the thinking prevalent in Americanist archaeology during the first several decades of the twentieth century:

> The division of the Glaze ware of Pecos into six chronologically sequent types is a very convenient and, superficially, satisfactory arrangement. For some time I was very proud of it, so much so, in fact, that I came to think and write about the types as if they were definite and describable entities. They are, of course, nothing of the sort, being merely useful cross-sections of a constantly changing cultural trait. Most types, in reality, grew one from the other by stages well-nigh imperceptible. My groupings therefore amount to a selection of six recognizable nodes of individuality; and a forcing into association with the most strongly marked or "peak" material of many actually older and younger transitional specimens.... This pottery did not stand still; through some three centuries it underwent a slow, usually subtle, but never ceasing metamorphosis. (Kidder, 1936a:xx)

Kidder's comments reveal the fundamental paradox: The conceived "slow, usually subtle, but never ceasing metamorphosis" of artifact forms through time was being monitored using typological "recognizable nodes of individuality." Kidder realized that his recognizable nodes masked significant variation, especially relative to older and younger "transitional" specimens. But rather than think about the epistemological and ontological implications of such transitional, or what were sometimes referred to as "hybrid," specimens, archaeologists attempted to sort specimens into types that measured small amounts of time. Temporal types built by extensional trial and error were bound eventually to become index fossils as their definitive criteria became so refined that the type units reflected ever smaller chunks of the time–space continuum. Such types were useful for purposes of keeping track of time, but as we will see, they took on a life of their own when used as icons for a particular culture or ethnic group.

Conceptions such as Kidder's (1936a) regarding types set many efforts at classification on a distinctive course. For example, in the Southwest, where superposition was first used so effectively and where dendrochronology provided absolute chronological information, definitions of historical types could be continuously refined in light of new information. This led to the creation of historical types with extremely narrow temporal distributions, in essence, index fossils (e.g., Breternitz, 1966). Their utility for analytical exercises such as seriation, which requires historical types with long temporal durations, was thus compromised. Further, artifacts used as index fossils, at least in part because of the fine-scale temporal resolution they provided, often were used to measure time discontinuously. Whether one begins with a conception of change as continuous or as discontinuous or one begins by measuring temporally sensitive phenomena in such a manner as to reflect continuity or to reflect discontinuity, the result is the same. The first of either pair will show time as a continuum, the second of either pair will show time as a set of discontinuous chunks. This, in our view, is the root of the modern debate between those subscribing to what has been characterized as the "Modern Synthesis," or the so-called gradualistic version of biological evolution, and those subscribing to the punctuated equilibrium version of biological evolution (e.g., Eldredge, 1982; Eldredge and Gould, 1977; Fortey, 1985).

Regardless of the epistemological problems involved, archaeologists have had success keeping track of time using index fossils, or types that occur over relatively small spans of time. All such markers usually begin as part of a type that spans a relatively large span of time, but through time they get whittled down in terms of the distinguishing characters used to delineate them. As variation gets sliced off, the resulting types, sometimes referred to as subtypes or varieties, begin to mark shorter and shorter time periods (Fig. 6.1). There are many examples of index fossils in Americanist archaeology. We focus on three that represent innovative thinking in terms of how index fossils are constructed and fairly represent how archaeologists have used such markers to slice up time and to correlate various archaeological manifestations in time.

FOLSOM AND CLOVIS POINTS

As we saw in Chapter 1, purported finds of early tools in North America were long invalidated on various grounds, but the breakthrough came in 1927 when workers from the Colorado Museum of Natural History recovered several small, fluted points in association with the remains of extinct bison (*Bison antiquus*) in an arroyo near Folsom, New Mexico (Figgins, 1927). The stratigraphic association was in a geological context that lay near the temporal border between the late Pleistocene and the early Holocene epochs, or about 11,000 years ago. Even in the 1920s, stratigraphers knew what the glacial-age boundary in the western United States looked like in terms of sediments and strata, and thus the age assessment of the bison kill site was not a shot in the dark. Importantly, the excavation crew left the points and bones in place and called in as witnesses several scientists whose reputations were above reproach: Kidder, paleontologist Barnum Brown of the American Museum of Natural History, and archaeologist Frank H. H. Roberts of the Smithsonian Institution. Everyone agreed that the artifacts and bones were contemporaneous. And Brown used biostratigraphic principles to pronounce the bison skeleton to be of late glacial age; that skeleton was morphologically similar

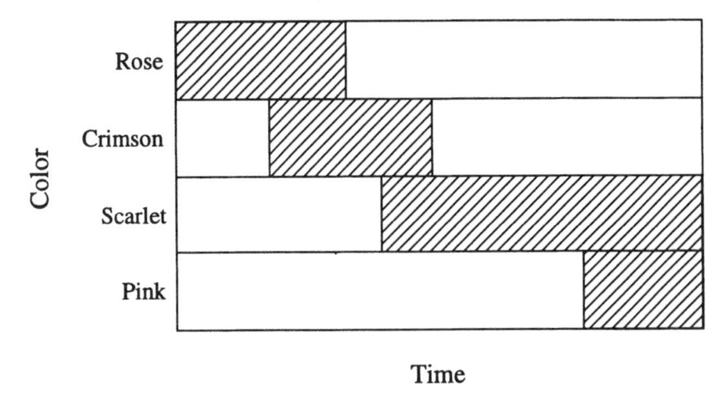

Figure 6.1. Diagram showing how an existing type can be divided into other types (often referred to as varieties), which then are used as index markers. Here a type carrying the undifferentiated characteristic "red" has been divided into four types on the basis of differentiation in shade. The new types have overlapping temporal distributions of varying degree.

to bison skeletons that had been removed from clearly glacial-age deposits elsewhere. Precisely such reasoning would shortly be used with respect to fluted points rather than bones.

The 19 projectile points unearthed at Folsom were easily recognizable because of the fine flaking and the presence of channel scars, or flutes, on both faces that resulted from the removal of long flakes. It was impossible to confuse this kind of point with more recent kinds, and it instantly became a chronological marker eventually referred to as the Folsom type (Fig. 1.1). In those early days no distinction was made between Folsom points and what later would be known as Clovis points (Fig. 2.7). Neither was it then recognized that Folsom points dated slightly later than Clovis points over most of the West and Midwest and that Clovis points would eventually be found associated with mammoths and still later with mastodons. That kind of evidence was soon found at Blackwater Draw in eastern New Mexico, where mammoth bones and Clovis points, what were labeled as "generalized Folsoms" or "Folsomoid," were found in strata beneath and separate from strata containing bison bones and "true Folsoms" (Howard, 1935). Such apparent discontinuities in point forms caused archaeologists to reexamine their collections of fluted points and to start subdividing them based on morphological characteristics other than simply the presence of fluting (see LeTourneau, 1998, for a detailed history).

Folsom points resemble Clovis points, especially reworked Clovis points, in that they both have concave bases and flutes on one or both faces, but there are a number of characteristics that since the late 1930s have been used to distinguish the two. Whereas Clovis "flutes" often were made by removing more than one flake from the base of the points, on Folsom points the flutes were made by removing one large flake [in a few cases sequential flakes were removed to deepen the flute (Bradley 1991:378)]. Folsom and Clovis points both have concave bases, but the bases of many Folsom points are flat as opposed to parabolic, the latter typical of Clovis bases. Folsom points also often exhibit a nipplelike projection in the middle of the concave base, which served as the striking platform for removal of the channel flakes. On other points, the nipple was removed after fluting. On some points the channel scar terminates in a hinge fracture just below the tip, but on other specimens, especially those that were reworked, the channel continues out to the tip. After a point was fluted, the lateral margins, those portions between the edge of the channel and the edge of the point, were pressure flaked, as was the tip.

Uncalibrated radiocarbon dates for Folsom points generally fall in the 8950–8500 BC range and those for Clovis points generally fall in the 9250–8950 BC range. If chronologist R. E. Taylor and his associates are correct, "the latest North American Clovis occupation predates the earliest occurrence of Folsom" (Taylor *et al.*, 1996:523). Clovis and Folsom points have become excellent index markers for the early portion of what generally is termed the Paleoindian period. Even before the Clovis type was created, fluted points in aggregate were excellent chronological markers, but their separation into a Clovis type and a Folsom type meant that even finer temporal control was possible. Just as with biostratigraphic observations, the recurrence of one type of point—Clovis—with mammoths and the recurrence of the other type—Folsom—with bison strengthened the associations first found at the Folsom and Blackwater Draw localities. By the late 1930s, archaeologists knew they had two good index types to use in dating the initial part of the North American archaeological sequence. Precise dating of Clovis and Folsom points had to await the advent of radiometric dating in the 1950s, but this did not hinder application of the index fossil concept.

The utility of fluted points, whether Folsom or Clovis, as index fossils that allow cross dating, that is, the temporal correlation of geographically separate archaeological manifestations, is unquestionable given modern control of their absolute time ranges. However, even such fine-scale temporal control as is presently available does not mean that all such correlations are sound. Prior to their dating via radiocarbon, it was clear that fluted points were late Pleistocene–early Holocene in age. They were the only such point style that clearly occupied the basal cultural strata, so it was sometimes thought that the archaeological record in an area lacking such points must not include the earliest temporal portion of the cultural record. For example, Douglas Osborne (1956) stated that he earlier had concluded that the archaeological record of Washington state must not extend back prior to about 6000 BC, because no fluted points had been recovered. In the mid-1950s, several fluted points were discovered in the state, and a radiocarbon date of about 6700 BC associated with stemmed projectile points was reported. Osborne (1956:44) changed his views and suggested that although the archaeological record of the state had a greater time depth than he had anticipated, fluted points probably occurred later in time there than in the Southwest as a result of the migration of fluted point peoples from the Southwest to the Northwest. Clovis points from central Washington state subsequently have been dated to approximately 9000 BC (Mehringer and Foit, 1990), so perhaps Osborne's suggestion is incorrect. His initial conclusion regarding the absence of fluted points displays a belief that a good index fossil such as is represented by fluted points will mark a particular part of the temporal continuum regardless of geographic position. His hypothesized relatively late age for such points in Washington is more sophisticated because it displays an awareness of the time-transgressive nature of index fossils that was lacking in his initial conclusion.

One still detects a sense of awe when a fluted point is found in an area where one has not previously been found. We suspect this is because such points, when uncritically conceived of as index fossils, are still often viewed much as Osborne viewed them in his initial conclusion: that they denote a particular small part of the temporal continuum over much if not all of North America. Thus, the discovery of a fluted projectile point on the northwestern coast of California was heralded as completing the empirical "documentation of the coast-to-coast distribution of this artifact form" (Simons *et al.*, 1985:260). This is a reasonable conclusion because it is based on the empirically documented distribution of a particular index fossil. Where the authors get into potential trouble is when they state that given their point's "overall stylistic similarity to other fluted points considered to be of late Pleistocene age, [it has] a projected maximum absolute age of about 11,000 BP" or 9000 BC (Simons *et al.*, 1985:265). Two sources of potential error exist here. First, why should mere similarity of form denote similar age, particularly when the point under consideration "does not appear to closely resemble 'classic' Clovis or Folsom point forms described from 11,500- to 10,000-year-old sites on the Great Plains and in the Southwest" (Simons *et al.*, 1985:265)? As we noted in the beginning of this chapter, such documented heritable continuity between compared specimens is unnecessary to biostratigraphic correlation, but it would certainly strengthen any inference of temporal similarity. Second, even after establishing heritable continuity between compared specimens, one must contend with the time-transgressive nature of index fossils before ascribing a fine-scale age to a specimen. Although we suspect Simons *et al.* (1985) are correct in their ascription, their example of how index fossils are used by archaeologists helps illustrate some of the potential pitfalls that await those who would use fluted points and other marker types uncritically.

GEORGE C. VAILLANT AND THE MEXICAN FORMATIVE

George Vaillant received his doctoral degree from Harvard in 1927, the same year Jesse Figgins was making the Folsom discovery. By that time, Vaillant had worked with two of the preeminent figures in Americanist archaeology—A. V.

Kidder at Pecos Pueblo in 1922 and George Reisner in Egypt in 1924—and had spent time working with the Carnegie Institution in the Yucatan region of Mexico. He joined the American Museum of Natural History in 1927, and the next year, at the direction of Clark Wissler and Clarence L. Hay, began working in the Valley of Mexico. The problem facing Vaillant was to place the "Archaic" culture identified by Manuel Gamio in the lowermost strata at Atzcapotzalco (Fig. 5.8) in time. Was it, as was widely accepted, a basal culture that had arisen in the Valley of Mexico and spread to other parts of the New Word, carrying with it pottery and agriculture, or was it simply one of several early cultures that had arisen simultaneously in Mesoamerica? Vaillant excavated at numerous sites, but we focus on only three: Zacatenco, El Arbolillo, and Ticomán. Today we know the first two sites date to the Middle Formative (Preclassic) period, or roughly 900–400 BC, and Ticomán dates slightly later.

Vaillant knew well the details of local archaeological research that had preceded his own (e.g., Gamio, 1913, 1924; Kroeber, 1925), and he devoted space in many of his reports to reviewing available data. As did his predecessors and contemporaries, Vaillant believed that at least some of the variation in pottery he recovered in large quantities was attributable to difference in age. Given his research goal, once in the field Vaillant looked for a site "possessing deep beds of débris" (Vaillant, 1930:18). The first site with such beds that he excavated was Zacatenco. Natural erosion and human activities had churned the deposit, but Vaillant focused on its overall thickness and depth (more than 3 m), noting in particular that

> Even if one had no sure geological demarcations of strata, at least with such depth one could arrange time periods from the different styles of artifacts found at various levels. If the rainy season washed the upper strata down over the lower in later times, it must have done the same each year in early times, so that although the original position of the débris might be altered, it would nevertheless preserve an equivalent deposition on its new bed. (Vaillant, 1930:20)

Importantly, Vaillant elaborated on the potential obfuscating factor of what came to be known as reversed stratigraphy:

> It is theoretically possible ... to have, through such agencies [as rainy season erosion], a reverse stratification, i.e., when the successive removal of strata by erosion presents the original layers reversed in the new accumulation. A pit dug into normal strata may produce this phenomenon in the excavated dirt.... The possibility of a reverse stratification must be considered, but, for several reasons it can only superficially affect the interpretation of the site. [Erosion of strata from topographically high to topographically low areas might produce a reversed stratigraphy, but] At some point on the slope one could detect the interlaminations of the strata and derive therefrom a true picture of past conditions. (Vaillant, 1930:20–21)

Given this cautious awareness of the potential trouble with stratified deposits, he indicated that

> Reliance had to be placed on the objects themselves and constant care exerted to extract them from their relative positions. Working against a hillside absolute depth meant little. The variability of the strata made it impossible to work by peeling of layers; and the material occurred more often in lenses than in prolonged depositions. To keep some relative control of the position of the objects, we moved into the deposits on a series of floors. Later, when the trench was opened completely, we could trace the débris lenses and fix the position. By recording the daily progress of excavation in section and plan it was subsequently quite easy to compare a digging level and the objects which it yielded on any given day with the actual deposition of the lenses, which could not be understood until after complete excavation of the trench. Except to begin a trench, no digging was ever done straight down. We worked against a vertical face in front, so that three faces, the front and the sides of the trench, were always exposed; and control was thereby established on dips in the strata. (Vaillant, 1930:22)

Vaillant used this technique to excavate a number of trenches at Zacatenco, and (1) having "arrived early in the work at a recognition of three main types of material, Early, Middle, and Late"; (2) failing to find continuous artifact-bearing strata segregated by "sterile layers"; and (3) recognizing redeposited materials, Vaillant (1930:28) was left with "the only alternative of deductive stratification on the most common occurrence of types of figurines and pottery." That is, horizontally separate strata in his various excavations were correlated using their artifact contents, a form of biostratigraphy or more correctly artifact stratigraphy (Table 6.1).

Vaillant's (1930:66–77) chronology of ceramics at Zacatenco aligned with those developed by earlier workers at nearby sites (e.g., Gamio, 1913; Kroeber, 1925). Like many culture historians, he cautioned that "Further studies must follow to corroborate the data acquired and refinements, both in excavation and typology, are essential to a more complete understanding of the problem of the cultural and chronological position of the early cultures in the Valley of Mexico" (Vaillant, 1930:66). His reports included exhaustive descriptions of the materials he recovered, with numerous photographs and drawings, typically arranged by material, type, and chronological position. His pottery types were largely based on decoration such as a "blue-white" type and a "red-on-black" type, and in this respect they were little different compositionally from what his contemporaries created. His typology of figurines, on the other hand, was unusual. Types were labeled with a capital letter, sometimes followed by a lower-case Roman numeral (Fig. 6.2).

In his initial report Vaillant (1930) had little to say about the reasoning behind this classification system, but in the volume summarizing the second field season, spent at Ticomán, Vaillant provided some insight into the purpose and reasoning behind his classification:

> The method evolved for the classification of pottery types from Zacatenco and Ticomán is not an absolute technical arrangement but a more or less eclectic grouping that, while following basal similarities in composition, at the same time lays stress on the stylistic

Table 6.1. Distribution of Figurines from George C. Vaillant's Excavations at El Arbolillo, Mexico, Arranged by Grouping Trench Cuts[a]

Figurine types	Trench D cuts I–IV	Trench B cuts I–III	Trench G cut I	Trench B cuts II–III	Trench G cuts II–III	Trench I cut I	Trench I cut II	Trench B cuts IV–VI	Trench I cut III	Trench G cut III	Maximum grouping percentage
A	5	0	1	0	0	2	0	0	2	0	80
B	8	5	1	0	0	2	0	0	0	0	
F	2	1	1	0	0	0	0	0	0	0	
F–C	0	0	0	0	2	0	1	0	0	0	80
B–C	0	1	2	3	1	0	0	0	0	0	
C1a	0	0	1	1	4	3	3	0	1	0	77
C1b	1	2	0	1	3	3	2	4	0	0	
C3c	0	0	0	0	2	0	0	1	1	0	
C3d	0	3	0	0	0	2	0	1	0	0	
C2	0	1	0	0	0	2	2	0	1	1	
C1–2	4	2	0	1	4	9	2	2	1	4	
C3a	0	1	0	2	1	0	0	3	2	3	65
C3b	0	0	0	0	2	0	0	0	1	2	
D	0	0	0	0	3	0	0	0	0	0	
Early F	0	1	0	0	0	0	0	0	1	0	
C5	0	0	0	0	0	0	0	0	1	1	

[a]From Vaillant (1935b).

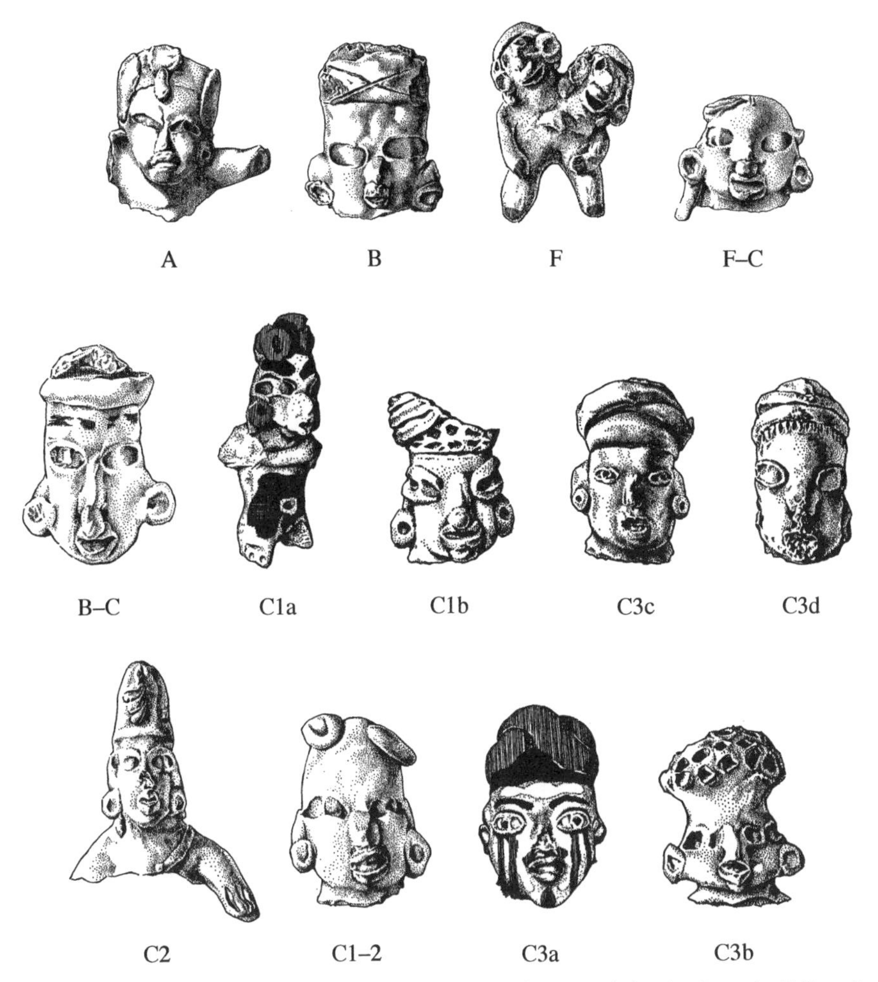

Figure 6.2. Figurine types devised by George C. Vaillant for use as index fossils in the Valley of Mexico (after Vaillant, 1935b).

vagaries which, resulting from the whims of fashion, reveal in consequence chronology. Figurines are sorted on a similar basis....

The bewildering number of designations for pottery styles and of letters to indicate figurine types need not unduly intimidate the student as they are meant purely to facilitate brevity and accuracy in exposition and reference. Eventually, when the groundwork for a chronology will have been completed, the student will be able to seize the main definitive types of each period and the elaborate preliminary nomenclature will be happily forgotten. Yet, in these first steps one must follow a fairly refined system of grouping material, lest one slur over details that might hold subsequent importance.

However, to simplify this detail as much as possible, we have labeled by the term,

> *time-bearer*, the more important chronological indications of the following digest of elements comprising [a] culture sequence. (Vaillant, 1931:210, emphasis added)

As was typical of Americanist archaeology in the 1920s and 1930s, types were constructed extensionally by trial and error, and thus could be modified in light of new material. Once types that allowed the measurement of time had been constructed, the details of the exploratory classification system could be dispensed with. Some types that allowed the measurement of time were ones that had rather limited temporal distributions, and thus became index fossils: Vaillant's "time-bearer" types. In his summary descriptions of the cultural materials associated with each of the periods in evidence at Zacatenco, Vaillant noted that some historical types tended to have unimodal frequency distributions through time, but he also noted that "new" types that appeared in the middle of the sequence "very likely entered from some other source. Thus there is the strong probability of the fusion of two peoples" (Vaillant, 1931:28). That is, types with no apparent evolutionary ancestral type must be culturally (not geologically) intrusive.

Excavations at Ticomán were meant to refine the chronology and to test this later notion of fusion. Vaillant collected 52 samples of sherds and figurines from stratigraphic proveniences and after studying them concluded that

> we were dealing with an autochthonous evolution. In other words, the development of material culture was self-contained and was subject neither to the amalgamation of foreign with indigenous elements [as at Zacatenco]. While each class and style of object went through change in the course of its history at Ticomán, there was really no simultaneous shift [in styles of both figurines and ceramic vessels] such as one would expect were fusions or dispersals of peoples involved. (Vaillant, 1931:251)

Vaillant's interpretive algorithm became commonplace in archaeology. On the one hand, abrupt shifts in the styles or types of pottery and/or figurines suggested immigration or at least a nonindigenous source, and thus perhaps diffusion or trade. Vaillant detected shifts in the styles of both ceramic vessels and figurines at Zacatenco, which reinforced the interpretation of the appearance of a new group of people rather than merely the diffusion of a few new ideas; the scale or magnitude of the change was critical to inferring whether diffusion or migration was the cause of change. On the other hand, at Ticomán the deposits were thin, apparently lacked "consecutive strata," and thus seemed to represent a single period. Thus, "as is so commonly the case in the history of a single people, there were no simultaneous shifts of the various styles of artifacts, so that a chronology for one class of object was not necessarily the same as for another" (Vaillant, 1931:329). That is, styles of figurines changed independently of styles of pottery through time. One could trace the *in situ* evolution of a culture by identifying stratigraphically superposed figurine styles that evolved one from another; in short, phyletic seriation denoted a lack of outside, or nonindigenous, influence. Vaillant's equation of particular sets of artifact styles with particular "peoples," "civilizations," or "tribal entities" was an archaeological mainstay by the 1930s. As had many of his contemporaries, Vaillant (1936:325) noted that one could not speculate "as to the identity and tribal affiliation of the [early pottery and figurine] makers," but this did not dissuade him from believing that "from the styles and types of their artifacts we can readily distinguish whatever sites they occupied." The last was simply biostratigraphy accomplished with artifacts rather than fossils; a particular artifact style found in different places denoted similar age, but also denoted a phylogenetic relation and heritable continuity.

Later field seasons saw Vaillant undertake excavations at other sites in the Valley of Mexico and a continued search for culture complexes. Some excavations were not clearly stratigraphic, and others used 50-cm-thick arbitrary levels. One important innovation in reporting was the plotting of particularly diagnostic styles of figurines—index fossils—on the drawings of stratigraphic profiles (e.g., Vaillant, 1935b; Vaillant and Vaillant, 1934). As well, the figurine typology was described in slightly more detail, and an attempt was made to legitimize it:

> Letters are used to designate major types and supplementary numbers in Roman numeration indicate the sub-types formed by minor stylistic variation [Fig. 5.3], sometimes ethnographically and sometimes chronologically significant. Considerable legitimate criticism may be brought to bear on this method of nomenclature as being too elaborate and too difficult to remember. But the use of regional or type site names would be even more confusing, for several types often occur at the same site. Thus the letter and numeral system of classification lends itself to exacter reference. The underlying idea is not to classify in an arbitrary manner, but to present to students the kinds of figurines found at different levels and at different places. At the same time, it is desirable to have a classification sufficiently elastic to include the expansions made necessary by fresh research. (Vaillant and Vaillant, 1934:24)

In his later report on excavations at El Arbolillo, Vaillant (1935b) repeated his earlier statements about the classification of figurines, but this time he added several critical details. In particular, he noted that specimens were grouped

> on the basis of the sum total of all their parts, plus their position in time and space.
>
> Under this system a group of figurines, restricted to a single site and a single period, would therefore present a greater variation in its constituent specimens, than would a group which could be subdivided into chronological divisions or regional styles. In this latter circumstance differences in the minutiae of detail would receive greater emphasis than the gross resemblance of the plastic technique as a whole. Thus [a] group of figurines which is widely distributed in time and space must be divided into a quantity of subtypes to distinguish these aspects, whereas [a] group which is confined to [one] culture phase exhibits no comparable range in local or chronological variations and does not require such subdivision. However, in some cases subdivisions were made when it seemed probable that future research would show this chronological and regional differentiation....
>
> In general the letters of the alphabet indicate the broad divisions of figurines according to technique and time. The numerals following indicate subdivisions on the basis of regional styles or a cross-cutting of a major group by time periods. Letters in small case following such numerals are the result of the discovery of time or regional variations in a type previously classified. (Vaillant, 1935b:190)

The basic structure of Vaillant's classification system seems to have been a logical predecessor of the type–variety system of the 1950s (e.g., Phillips, 1958, 1970; Wheat *et al.*, 1958). The system was devised to "preserve data of potential historical worth, until sufficient information is amassed for secure recognition of the interplay of culture in Mexico. Then a simpler system can be confidently adopted, with a nomenclature indicative of tribe and period" (Vaillant, 1935b: 190). Similar but less lengthy comments were made regarding the classification of ceramic vessels (Vaillant, 1935b:190–191). But Vaillant never once provided an explicit definition or listed the definitive characteristics of any of his figurine or vessel types. Instead he provided elaborate descriptions of many of the members of each type. Thus his mentor, Alfred Tozzer (1937:340), complained that he had tried to apply Vaillant's classification, but the latter "had to come to [the former's] rescue."

By the early 1930s Vaillant's figurine types were tightly, if not explicitly, defined ideational units that could display distributions over space and through time. Vaillant's (1935b:159) chronological periods were termed "figurine periods" or "figurine phases," clear evidence that time was being measured in terms of the presence or absence of index fossils. Further, Vaillant (1935b:189), at the end of his research, was explicit about the fact that figurines and pottery vessels comprised two divisions of ceramic art, "each susceptible of close classification and capable of use as a control for the historical analysis of the other." That is, because each of the two divisions had a unique evolutionary and developmental history, the chronological implications of one could serve as a check on the chronological implications of the other when types of the two categories were found associated. This was innovative.

In his two semipopular summations of his excavations, Vaillant (1936:324–326) spoke of figurine types, "usually female," having perhaps "been used as votive objects or as household images of saints," and of "ornamental vessels [serving] in rituals," of "peoples" as equivalent to cultures or periods, of different "tribes" being represented by different ceramic styles. Likewise, Vaillant (1937: 313) spoke of distinct "culture levels [that differed] from each other in the form and decoration of their pottery, in the artistic styles of their stone and clay sculptures, and in their architecture. Through the study of the strata in the rubbish heaps, minor time stages can be distinguished within each culture group." Thus, while his empirical evidence suggested that culture change was gradual and continuous, he leaned more strongly toward viewing time as discontinuous, even suggesting that his strata might reflect 52-year-long cultural periods, corresponding to cycles in the Aztec calendar (Vaillant, 1937).

These musings were nothing but pure speculation, but they were typical of the period, given the desire on the part of archaeologists to be anthropological. More importantly, Vaillant's speculations were outgrowths of how time was being measured, and hence of how change was being perceived. Vaillant had, over a multiyear project, refined his ceramic vessel and figurine type definitions, by trial and error, to the point where many of them could serve as index fossils, what he termed "time markers." These allowed him to correlate various archaeological materials from different sites with one another, and they also allowed him to erect a regional temporal sequence because various of them were found in stratigraphically superposed positions relative to one another. That the correlations and sequences in fact measured the passage of time was indicated by the recurrent patterns he detected. This was good biostratigraphic procedure. When Vaillant used his time markers as icons for particular ethnic groups and time periods, he was measuring time discontinuously, and change could only occur at the boundaries between tribes and periods. There is nothing inherently wrong with such measurement, although as we noted at the end of Chapter 5 it masks variation. Not surprisingly, most index fossils defined by paleontologists come from superposed strata, nor is it surprising that artifacts used as index markers also come from superposed strata. We pick and choose a distinctive fossil or artifact form that is easily recognizable and that is earlier or later stratigraphically than other fossils or artifacts of different forms that, if sufficiently distinct, themselves might serve as index fossils. We must be ever cognizant, however, of how this basic procedure tints our perceptions of change. One person who was quite aware of this worked in the Southeast, and we turn to his work as our last example of cross dating and the use of index fossils.

JAMES A. FORD AND THE LOWER MISSISSIPPI VALLEY

As Vaillant was completing his excavations in the Valley of Mexico, a young James A. Ford was setting out on a career that would eventually vault him to the forefront of Southeastern archaeology (see O'Brien and Lyman, 1998, for a detailed history of Ford's career). Whereas Vaillant and other Americanists had used index fossils to solve local problems, Ford took them to a new level by using them to construct a regional chronology. He began working in the Southeast in the late 1920s while still a high-school student. Over the next few years he was hired by Smithsonian archaeologists Henry B. Collins, who was working in southwestern Mississippi, and Frank M. Setzler, who was working in northeastern Louisiana. Collins, along with several colleagues, spoke at the seminal *Conference on Southern Pre-History* sponsored by the National Research Council and held in December 1932 in Birmingham, Alabama, a meeting Ford attended. Collins (1932:40) described how the geographic distribution of historic pottery types in Mississippi aligned with the distribution of various ethnographically documented tribes, and he emphasized the importance of the direct historical approach to solving chronological problems—that is, starting at a known point in time (the historical period) and working back in time into the unknown based implicitly on a notion of heritable continuity of artifact form (see Chapter 3).

Based on what he heard at the conference, together with what he had learned from Collins and Setzler, Ford developed three tenets that guided his work: (1) cultural change and development was gradual and undeniable; (2) the archaeological measurement of time was of utmost importance; and (3) stratigraphy could help solve the chronology problem. Early on, Ford was explicit about his views on culture, culture change, and how to document such change:

> Culture is in reality a set of ideas as to how things should be done and made. It is in a continuous state of evolutionary change since it is constantly influenced both by inventions from within and the introduction of new ideas from without the group.... All [artifacts] were subject to the principle of constant change, hence those on any one site are *more or less* peculiar to the time that produced them. (Ford, 1935a:9)[1]

In short, if particular kinds or types of artifacts could be tied down chronologically, they would be excellent index fossils.

Initially, Ford (1935b:1–2) mimicked, without citing them, the efforts of Kidder and Nelson and identified seven "decoration complexes," or groups "of pottery decorations characteristic of an area at a definite period of time" (Fig. 6.3). Each complex consisted of several pottery categories, generally termed "types" by Ford. He established the temporal relations among three of the complexes by excavating the Peck Village site in Catahoula Parish, Louisiana. Ford excavated in arbitrary levels, the thicknesses of which were determined by the amount of material collected; each level contained "an appreciable amount of material" (Ford, 1935b:6). As we noted in Chapter 5, given his belief that culture change was continuous and gradual, it is not surprising that he used such an excavation technique. Any vertical units are depositional accidents or accidents of where one locates a datum plane to which arbitrary levels are tied, given that culture change is continuous through the vertical sedimentary record. Ford knew that one needed sufficiently large samples to monitor culture change without fear of sample size effects, but they did not have to come from natural strata or from artificial units of equal thickness.

In the Peck Village report, Ford (1935c) presented his spatial–temporal ordering of decoration complexes and then demonstrated how he arrived at the ordering. Because of its geographic location, Peck Village exhibited little pottery carrying the decorations evident on pottery from prehistoric sites in the western and northern parts of the study area (Fig. 6.3), but it did contain a large number of prehistoric-period incised "Marksville complex" sherds. What made the presence of those sherds at Peck Village so important was their stratigraphic position: The separation was not perfect, but there was clear indication that coming up through

[1]Ford's early work on chronology in the Lower Mississippi Alluvial Valley appears in O'Brien and Lyman (1999).

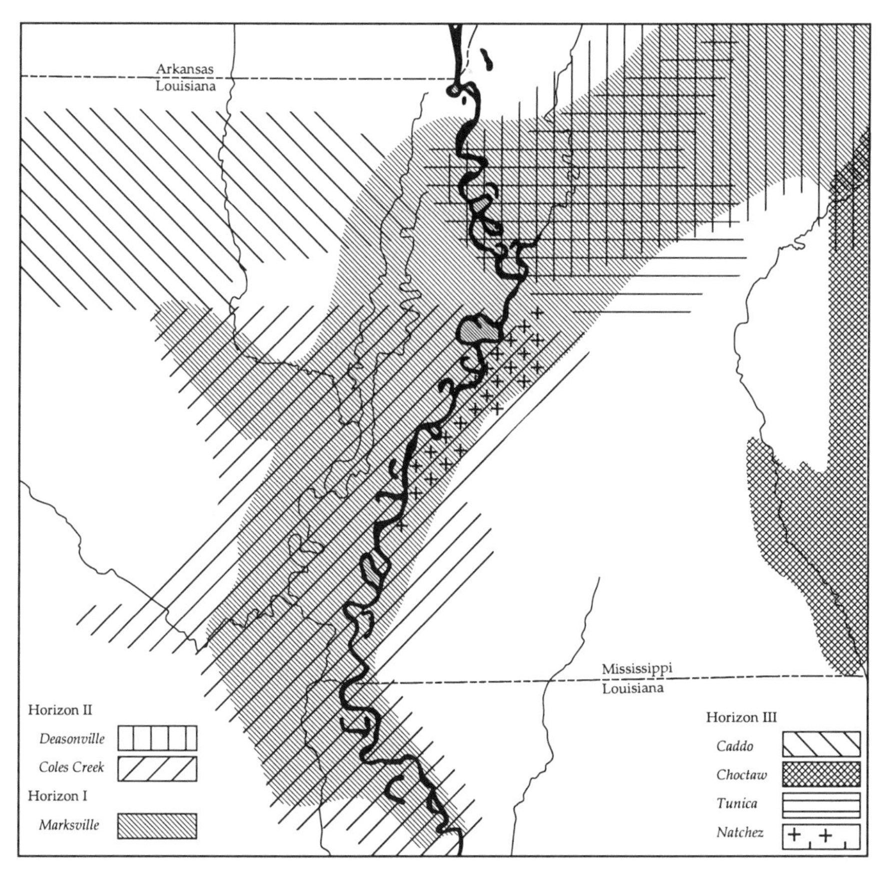

Figure 6.3. Map of northeastern Louisiana and southwestern Mississippi produced by James A. Ford showing his proposed distribution of seven pottery complexes (after Ford, 1935b).

the levels of the site, "marker type" sherds of the Marksville complex were gradually being replaced by sherds of the prehistoric Coles Creek complex.

To Ford, the Marksville complex was the basement of the Louisiana–Mississippi pottery sequence. At Peck Village, sherds of that complex were replaced by sherds of the Coles Creek complex, whereas at the prehistoric Deasonville site in western Mississippi that Collins (1927) had excavated, the Deasonville complex had replaced the Marksville complex. The pottery complexes of four historical period groups had in turn replaced the Deasonville and Coles Creek complexes in each of the two subregions—the western and northern on one hand and the eastern and southern on the other—though at different times (Fig. 6.4). Ford (1935c:23–24) noted that

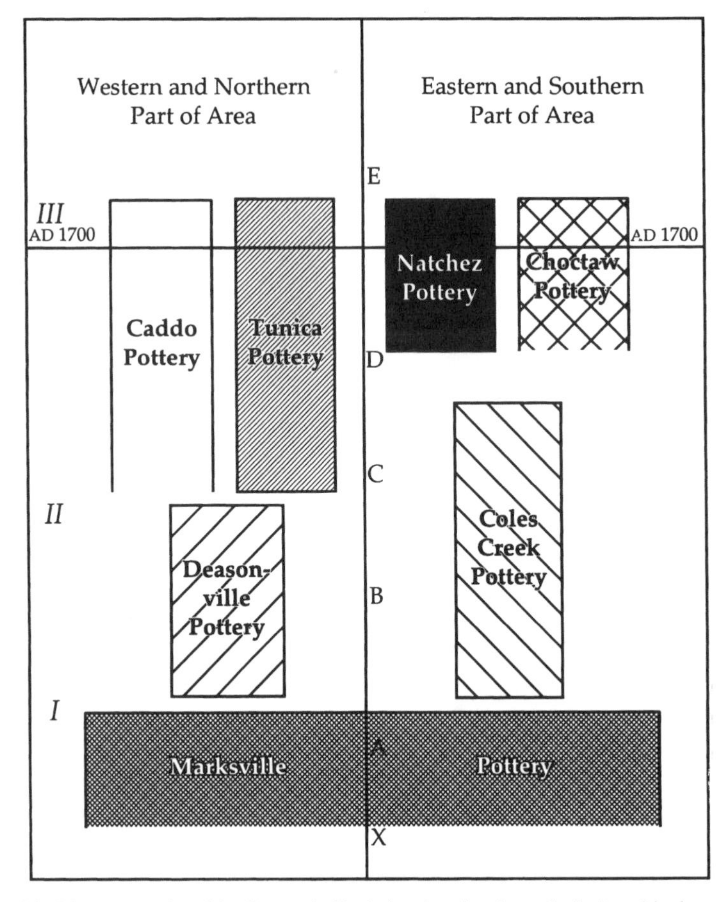

Figure 6.4. Diagram produced by James A. Ford showing the chronological positioning of seven ceramic complexes from northeastern Louisiana and southwestern Mississippi. Here Ford took the seven complexes shown in Fig. 6.3 and arranged them into two sets: those in the western and northern half of the area and those in the eastern and southern half. Points X–E are arbitrary points along the vertical time line. Note that both evolutionary trajectories sprang from a common Marksville base (after Ford, 1935b).

> The most important implication of the Peck Village situation is that with the passage of time, while deposition of the midden was in progress, the ceramic art of the inhabitants was slowly changing from decorations consisting of wide, deep, closely-spaced lines forming curvilinear and angular designs ... and curving bands of rouletting enclosed by wide deep lines ... to decorations formed with overhanging lines which usually encircled the vessels parallel to the rim ... and curvilinear lines with which triangular punctates were employed.... As there is no apparent typological connection between the dominant decorations of the two complexes, the change in ceramic art probably is not

> the result of local evolution of Coles Creek out of Marksville, but rather a *replacement*. The Coles Creek decorations appear fully developed although in all cases they are not as specialized or as neatly executed as at many sites of the pure complex. Some of the types which form a minor proportion of the Coles Creek Complex do seem to be typologically related to Marksville. They probably resulted from the gradual replacement of complexes, indicated by the smooth changes in proportion in the graphs. This may have allowed a certain amount of Marksville to be absorbed into the Coles Creek.

Here, Ford was attempting to explain the change from one decoration complex to another, and although he did not specify a particular mechanism for the change, in a slightly later monograph he indicated that he viewed replacement as occurring "either through gradual infiltration or conquest" (Ford, 1936a:5). There was little typological connection between the dominant decorative types—his marker types—making up the complexes, which suggested to him that Coles Creek replaced Marksville. In other words, the lack of typological connection between the dominant types of the two sequent complexes suggested the absence of a direct phyletic connection or heritable continuity between the types. But he also indicated that some of the rare, or nonmarker, Coles Creek decoration types seemed to be typologically related to Marksville. Thus there was a direct phyletic connection between the less dominant types. Ford believed that the decorations executed on the Coles Creek pottery from Peck Village were not as fully developed as those on pottery from sites that contained only Coles Creek complex pottery. This suggested that some of the Marksville designs had slowly been absorbed into Coles Creek. Thus the replacement of Marksville decoration by Coles Creek decoration was not complete because some vestiges of the former, probably at the scale of attribute rather than of type, remained in what he recognized as the later Coles Creek complex, but in muted or modified form.

Ford, like virtually all of his contemporaries, wanted to discuss the history of ancient cultures and different groups of people (Ford, 1935a), both of which are ethnographic units. Thus his archaeological units were warranted by the commonsense notion that they represented ancient and distinct ethnic units, just as the historic ethnic units variously termed Choctaw, Tunica, and the like could be distinguished by their ceramic decoration complexes. Ford's discussion of pottery design complexes was derived from his view that culture constantly changes. But he stumbled, like his contemporaries, because his complexes measured that change as discontinuous chunks, and those chunks—manifest as decoration complexes, the dominant types of which served as index fossils—were thought to represent, potentially at least, distinct cultures or peoples. Ford failed to explicitly acknowledge that the chunks were merely arbitrary units of measurement in the same way inches or grams are arbitrary units. Thus the kind of culture change he studied at Peck Village was one tied to the mechanisms of replacement or absorption, ethnologically documented mechanisms of culture change thought to be visible archaeologically. Replacement denoted the gradual cessation of one complex, a set of index types, and the initiation of another, and absorption denoted the continuation of a type across sequent complexes, a direct phyletic continuity of a decorative theme, but with modifications from outside. Again, these were interpretations much like those made by many of Ford's predecessors.

Ford, like many of his contemporaries, believed that pottery comprised the " 'key fossils' of culture" (Ford, 1936b:104). He noted that all "pottery decorations found at the Peck site were classified according to an index" that had been developed to study the large collections of sherds from the surfaces of sites in Louisiana and Mississippi (Ford, 1935c:8). Ford stated that even prior to implementing that index system of classification, he had considered and abandoned other methods:

> Noteworthy among the discarded methods is a morphological, biological-like arrangement of decorations into orders, suborders, families, etc. This proved unsatisfactory because of the extreme flexibility of pottery types, as well as doubt as to their generic [in the sense of a biological genus; thus, phylogenetic] relationships. The frequent migration of decoration elements from what seemed to be their native types was of significance, and could not be indicated by the method. Zoological classification is not embarrassed by such anomalies as would result from the frequent crossing of different species. (Ford, 1936a:17–18)

It is now abundantly clear that biological species do often interbreed, a fact over which biologists are not embarrassed at all.

There are important points in the above quote concerning a biological-like system of classification. Ford's comment that types contain "flexibility" suggests that he was not imposing order on the sherds by using previously defined ideational units. Rather, he was extracting type descriptions from the sherds themselves by placing all similar sherds in a pile and then deriving from that pile what an average or typical specimen looked like. This accounts for how Ford was able to determine which "types" were diagnostic of particular historical period tribes. Thus his types were flexible; their definitive criteria could change as specimens were added to or subtracted from a pile. As analytical constructs for measuring chronology, they were trial-and-error formulations. Although the types were constructs of the archaeologist, they also were something real that could "migrate" and "crossbreed" (see above quote). The problem was that Ford failed to clearly distinguish among the sherds (the stuff) he sought to classify, the units (types) into which he sorted the sherds, and the interpretations he rendered from the types.

Ford (1935c:8) referred to his system as "merely a list of decoration types," noting that as "distinct decorations were encountered in the collections in sufficient numerical quantity and areal distribution to permit their acceptance as a type, they were illustrated on an index card." In other words, the actual types he used were decoration based, but Ford recognized types only if they were represented by enough sherds and if they occurred across enough space to preclude them from being idiosyncrasies. Thus Ford's types were analytical units useful for measuring space and time if they occurred in superposed sequences and/or their frequencies fluctuated monotonically through time.

Ford (1935c:10) made a very perceptive statement when he claimed that

> The limitations and crudity of this means of classifying potsherd decoration are obvious. Increasing understanding of the chronology will doubtless demonstrate the stages of southeastern ceramic evolution and make possible a more analytical classification, which at present promises to be the result of rather than the means to prehistoric chronology.

This point, that the analytical classification was the result of chronology rather than a tool to measure or document it, underscores the fact that Ford constructed types by enumerating the characteristics of a set of objects rather than by identifying objects as belonging to a particular ideational unit.

Key to Ford's analysis of chronological change at Peck Village was the vertical distribution of what he termed "complex markers":

> From observations made on a number of village site collections in the lower Mississippi Valley, it has been noted that although each of the several decoration complexes of the area include a number of different and often unrelated types which appear at the various villages typifying any one complex, there are one or two small groups of closely related types peculiar to each complex that statistically dominate to a marked degree. These decorations must be considered as the most typical of their complex, and from the role they play serve as "complex markers." (Ford, 1935c:21)

In short, complex markers—"marker types"—were pottery types that could be employed as index fossils in the usual sense of the term. Marker types for the Deasonville complex are shown in Fig. 2.4, and those for the Coles Creek complex are shown in Fig. 6.5. Hence, for Ford a multicomplex site (a site, for example, containing sherds of both the Marksville and Coles Creek complexes) was identified as such only if it exhibited marker types representing more than one complex. Ford used the marker types, and only the marker types, to demonstrate the replacement of Marksville sherds by Coles Creek sherds at Peck Village.

Although sherds of other complexes were recovered from the site, Ford (1935c:22–23) noted that, particularly relative to Deasonville complex types, sherds

> are not present in sufficient numbers in the collections to show definitely the relation in which they stand to this indicated superposition of complexes.... [T]he Peck Village site lies well within the area covered by the Coles Creek Complex, and is a hundred miles south of the Yazoo River Basin where typical Deasonville sites are found. Sherds of this complex occurring at Peck site probably are the result of either trade or influence from the Deasonville area.

Ford's graphs of proportions of marker types indicate that Deasonville complex sherds were present in varying amounts throughout the sequence. But these and sherds of other, unmentioned complexes were not distinguished in his summary graph (Fig. 6.6) or in his discussion, since it was the overall waxing and waning

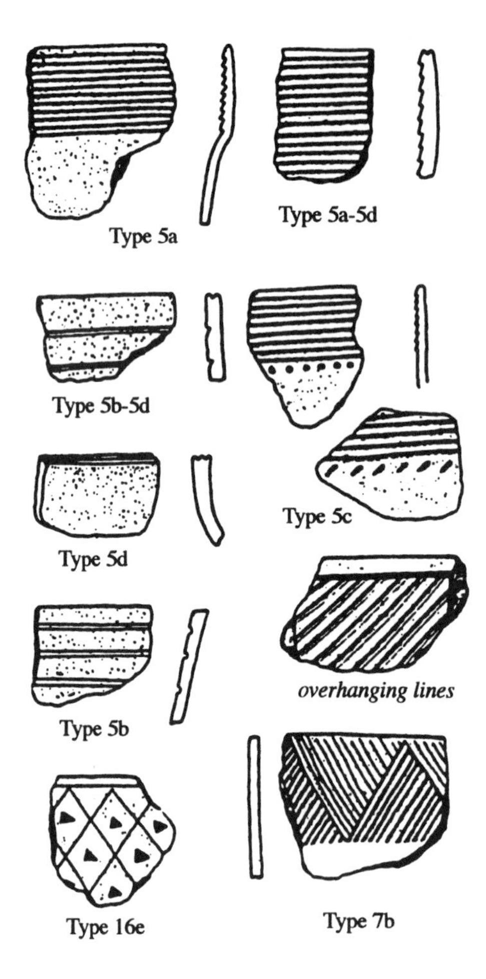

Figure 6.5. James A. Ford's illustration of designs represented on pottery marker types in the Coles Creek complex of eastern Louisiana (from Ford, 1935b).

only of the marker types in which he was interested. Such were "one element of culture [pottery] subject to rapid change in form" (Ford, 1935a) that allowed him to build a chronology and to correlate similar assemblages of pottery across large areas. Ford (1936a) did just that the next year when he examined the pottery he had recovered during his survey of 103 sites in Louisiana and Mississippi (Fig. 6.7).

After he published the Peck Village study, Ford abandoned the index system of type construction, noting that it was highly subjective. He claimed that it "was only semi-systematic, was non-analytical, was meaningless unless memorized in detail, and was not capable of logical expansion" (Ford, 1936a:18). Ford wanted something more manageable, replicable, and useful for measuring chronological

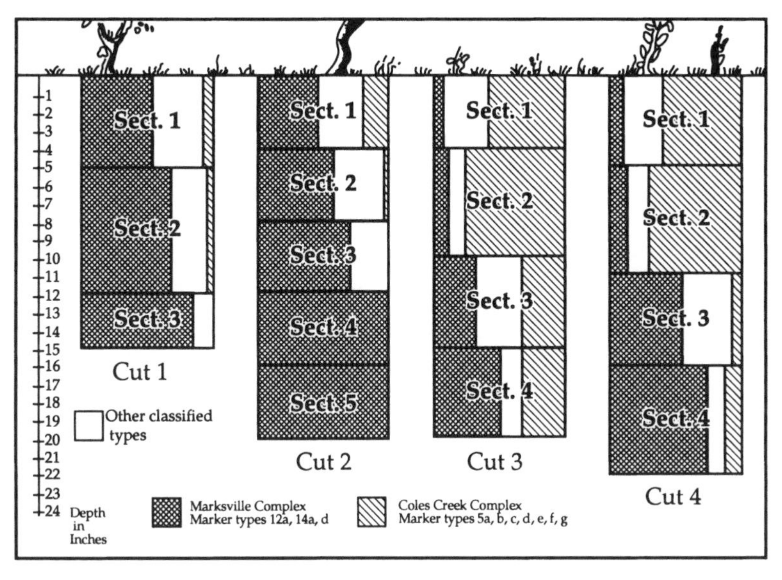

Figure 6.6. James A. Ford's depiction of percentage stratigraphy at Peck Village, Catahoula Parish, Louisiana, using pottery marker types (after Ford, 1935c).

variation. His new system was based strictly on decoration, which Ford viewed as comprising two "components": *motif*, the "plan of the decoration" (e.g., herring-bone pattern), and *elements*, the "means to express the motif" (e.g., incised lines) (Ford, 1936a:19). Ford's choice of the word "component" was unfortunate because this term was coming to have a rather different meaning in Americanist archaeology (e.g., Gladwin, 1936; McKern, 1934, 1937) than the "constituent part or feature" Ford meant to denote. The two components (we prefer the term *dimension* and use it hereafter) became the cornerstone of the new system.

Ford's units were ideational, that is, the necessary and sufficient conditions for membership were stated explicitly, and what is more, they were not extracted from the sherds but rather were used to identify the set of attributes displayed by the sherds. Types were particular combinations of attributes that were immutable, atemporal, and aspatial, that is, type A exists for all time and all space (just like an inch or a gram) as opposed to being time- and space-bound elements that were inherent in Philip Phillips's (1958, 1970) type–variety system (Chapter 2). Ford's system allowed him to sum the frequency of occurrence of any attribute, whether a particular motif or an element, as well as the frequency of any class, that is, the frequency of the actual representatives of a particular class.

Ford's classification system was, as Jon Gibson (1982:265) suggested, "the most sensitive and rigorous classification scheme to be used in the Lower Missis-

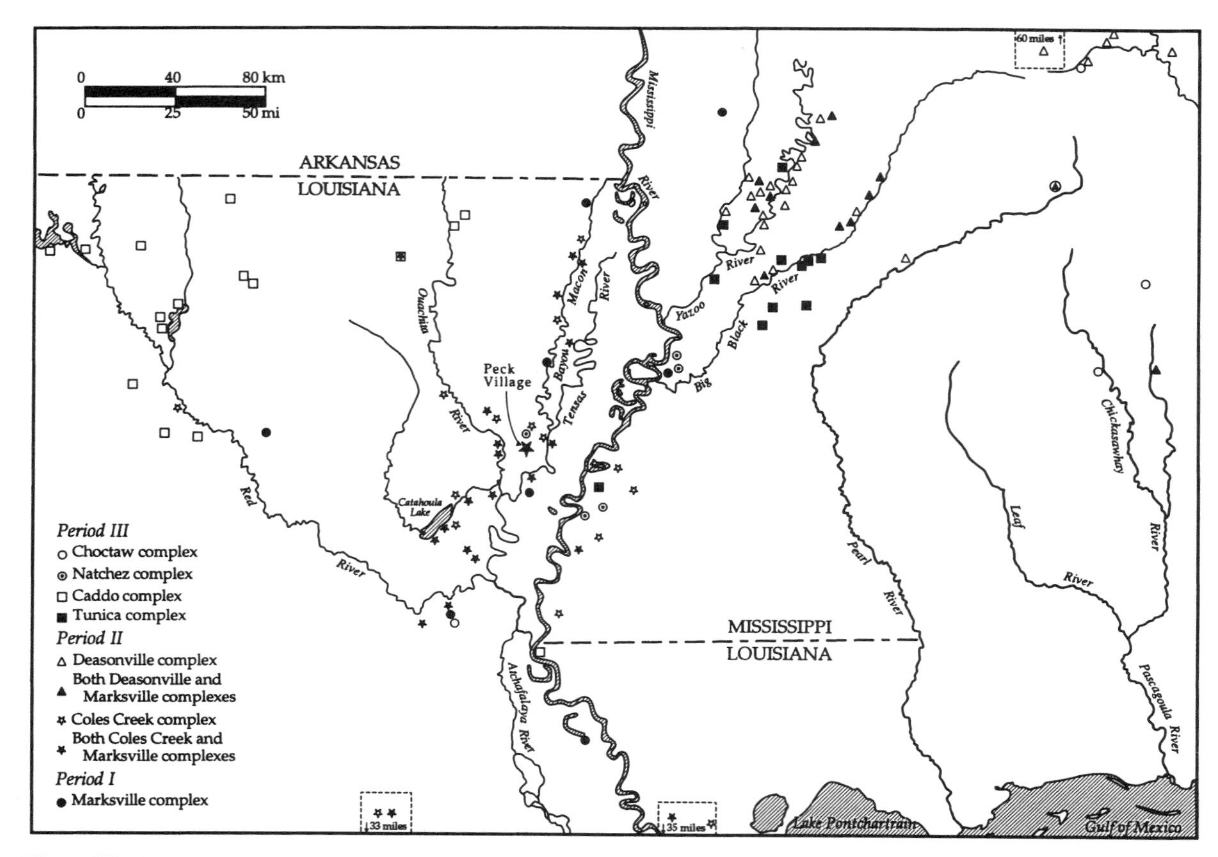

Figure 6.7. Map showing locations of James A. Ford's surface-collected sites in southwestern Mississippi and northeastern Louisiana (after Ford, 1936a).

sippi Valley." The system was based on objective criteria, an objectivity derived from direct definition of attributes and attribute combinations that were "uncontaminated by notions, suspected or stratigraphically demonstrated, of chronological order or cultural relationships" (Gibson, 1982:266). Although Ford did select attributes that he believed, based on previous experience, would allow the measurement of time and space, the classification was imposed on rather than extracted from the sherds. The system also was infinitely expandable. If, for example, new attributes were discovered during subsequent analysis, they could be accommodated simply by adding them under the appropriate component or dimension. This made them excellent index fossils. The net result of Ford's work in 1935–1936 was the production of a system of categorization of sherds that resulted in the creation of classes, the distributions of which could be plotted in time and space. The classes were ad hoc units because they were built by trial and error rather than being theoretically informed constructs, but they were ideational units nonetheless. Ford stated what the necessary and sufficient conditions were for unit membership and held to the system faithfully during the classification of specimens.

Using the marker types he had identified for each decoration complex, Ford sorted the collections of sherds from 103 sites into one of seven decoration complexes. His sorting has been referred to as a seriation but a more accurate description of it is artifact (bio)stratigraphy for purposes of cross dating because he did not order the collections within each complex. Rather, he merely determined the most abundant marker types in each collection, then noted which decoration complex was most frequently represented by the marker types, and finally placed each collection in its appropriate decoration complex. We calculated the average relative abundance of marker types per site-specific collection for each of the seven decoration complexes from Ford's (1936a:Fig. 1) data. Those averages are summarized in Table 6.2 and, although a crude measure of the accuracy of Ford's sorting procedure, they indicate that Ford's arrangement was reasonable. Sites are placed in the decoration complex for which they have the greatest abundance of marker types, though marker types distinctive of other decoration complexes are present in many sites.

Once his sites had been assigned to a decoration complex, knowing where each site was located allowed Ford to map the geographic distribution of each decoration complex. Ford referred to his method of chronologically arranging sites as " 'complex linking'—connecting time horizons by the overlapping of complexes occupying neighboring areas." This is remarkably similar to the concepts of *horizon* and *horizon style* discussed a decade later by Kroeber (1944) and Willey (1945) with reference to South American prehistory; we return to these concepts below.

Ford also realized that "mixing" of sherds of different decoration complexes could result from several factors. He assumed trade and diffusion had created the mixture of complexes seen in the 103 sherd assemblages he analyzed. He also was

Table 6.2. Summary of James A. Ford's Cross Dating of 103 Sites from Mississippi and Louisiana[a]

Decoration complex	Choctaw (1)[b]	Natchez (1)	Caddo (4)	Tunica (3)	Deasonville (2)	Coles Creek (5)	Marksville (3)
Choctaw (4)[c]	80[d]	0	0	+[e]	11	+	4
Natchez (3)	+	50	+	7	1	3	0
Caddo (13)	0	+	33	0	+	10	2
Tunica (5)	+	+	4	28	4	13	0
Deasonville (35)	0	0	+	1	57	12	7
Coles Creek (36)	0	+	+	+	9	32	15
Marksville (7)	0	+	0	+	2	1	71

[a]From Ford (1936a).
[b]Number of marker types.
[c]Number of sites assigned to the decoration complex.
[d]Values are average relative abundances of marker types per site in a decoration complex.
[e]Less than 1%.

well aware that reoccupation of a site could result in the mixing of decoration complexes. Reoccupation, that is, occupation after a preceding abandonment, was possible, but "it would be unlikely that a succeeding people should select the exact [previously occupied] habitation spot for their use. If the old locality had been intentionally reoccupied, the odds are that the dumps of the succeeding group would be located near but not precisely on those of the original inhabitants" (Ford, 1936a:255). In short, mixing of complexes resulting from reoccupation was unlikely. Mixing of complexes, Ford believed, was more likely the result of continuous occupation:

> It seems more reasonable to suppose that sites on which apparently subsequent complexes are mixed were either settled in the time of the older and were occupied on into the time of the following complex; or that the villages were inhabited during a period of transition from one complex to the other.... If the art styles changed without disturbance of the population, it is reasonable to expect that there should be such a period. (Ford 1936a:255–256)

Note that, apparently, even long-term continuous occupation was rare, at least in Ford's mind, because most sites did not contain thick accumulations of artifacts. The few that did have such accumulations contained, according to Ford, "transitional period" assemblages. To demonstrate that there indeed were such transitional periods, Ford (1936a:262, 268) discussed "certain decoration types which suggest that they are the results of an evolutionary trend which runs through two or more of the subsequent complexes," being careful to point out that such continuation "does not imply that this evolutionary process occurred in the local geographical area. In most cases it is more likely that the evidence is a reflection of the process taking place in some nearby territory." Here Ford was suggesting that particular attribute states of decoration types originated in and diffused from one area ("territory") to another, where they subsequently were incorporated into the local decoration complex by taking the place of or resulting in the modification of an attribute state of a vessel (Fig. 6.8). He noted, for example, how one type was found throughout the sequence of complexes, but it "also took on, in each complex, the features [read attributes] peculiar to that complex" (Ford, 1936a:263).

Ford's (1936a:263) "lines of development" provided historical links between

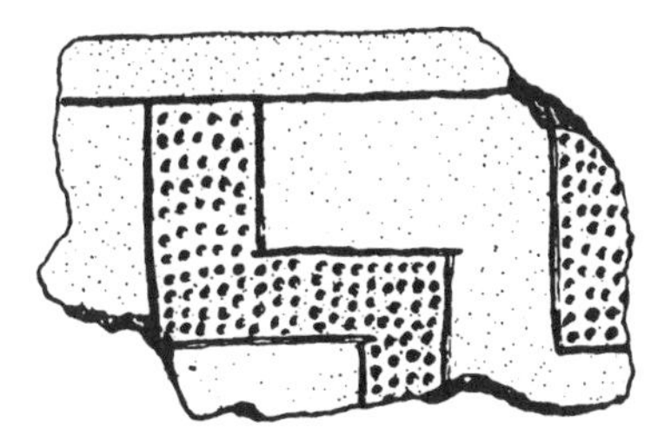

Type 32; $\frac{21}{71}$; 2, Caddo Complex
Period III

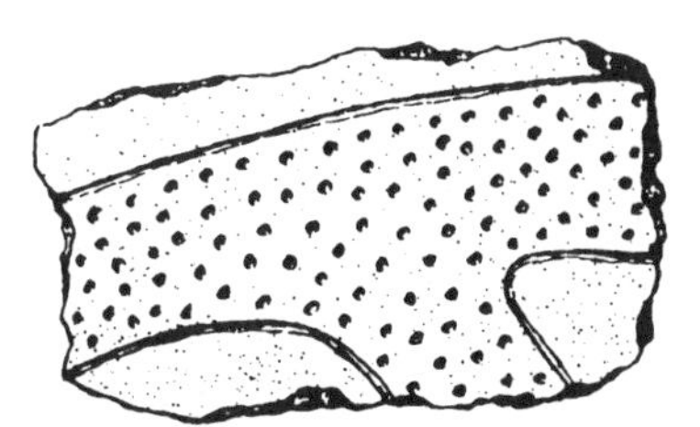

Type 31; $\frac{21}{71}$; 2, Deasonville complex
Period II

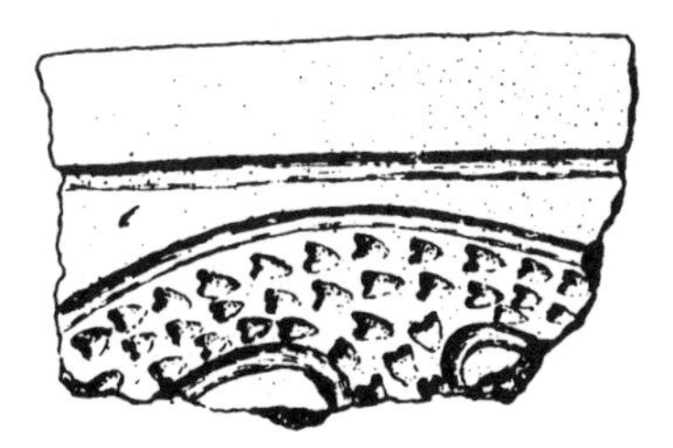

Type 31; $\frac{23}{73}$; 2, Marksville complex
Period I

Figure 6.8. James A. Ford's depiction of the possible evolution of pottery decorations from type 31;23/73;2 to type 32;21/71;2, based on samples from Louisiana (from Ford, 1936a).

and suggested evolutionary trends—heritable continuity—through two or more sequential complexes and involved the notion of phyletic evolution of one form into another via gradual alteration. In the cases he cited, alteration was at the scale of one or more attributes of a type. This was, of course, the phyletic seriation used by A. V. Kidder 20 years earlier and by W. M. Flinders Petrie and John Evans in Europe still earlier (Chapter 3). Ford's decoration types evolved as a result of influences from outside, an idea perfectly in concert with and probably derived from the then-prevalent notion that it was unlikely that a "complicated technique of decoration" could be independently invented in two distinct areas at approximately the same time (Setzler, 1933:153). New decoration complexes—aggregates of types—evident in an area thus represented the replacement of one culture by another, the same argument Vaillant made based on his work in Mexico. In both cases, the former at the scale of what we would call a discrete object and the latter at the scale of a set of multiple types, *in situ* evolution via invention was largely precluded because of Ford's view of culture change. Within a given culture, change was continuous and it usually was gradual. For Ford, this meant phyletic evolution was perceivable at the scale of attribute. Change in types was at the scale of discrete object and thus involved a combination of attributes.

The lack of apparent phyletic ties between types construed as sets of attributes and representing different time periods simply did not fit Ford's model of gradual and continuous change within a culture. There was no obvious cause for internal innovation or invention at such a scale, and this strengthened the notion that the source of innovation must be external. Apparent phyletic ties between one or two attributes suggested only a minor internal source of change that Ford's model could accommodate. At the scale of decoration complex, the major source of innovation must be external. Thus Ford (1936a:270) ended his monograph on the surface collections by noting that "Even with this modest beginning there is quite a temptation to see a story of ancient movements of people and cultural forces in the local region with ramifications spread over much of the eastern United States." These interpretations and their attendant problems aside, what Ford had done was to use index fossils or marker types to correlate a plethora of ceramic collections and sort them into several distinct spatiotemporal units termed *decoration complexes*. Pretty impressive for a person who at the time that he wrote the surface collection monograph had not yet completed his undergraduate education.

MEASURING TIME DISCONTINUOUSLY

By 1940, the use of artifact marker types was widespread. Archaeologists realized that finer and finer units of time could be measured using more tightly defined artifact types, so they reduced the variation in existing types to form such units. The marker types that resulted measured time discontinuously, though the intervals comprising the chunks were small. The problem was that marker types began to serve as indicators of particular cultures. Certainly such use is evident in efforts to make sense of the spatiotemporal distribution of fluted points, in Ford's work in Louisiana and Mississippi, and in Vaillant's work in Mexico. Once superposed strata were equated with cultures, it was natural that stratigraphy would be used to trace cultural succession and that index fossils were used for cross-correlating cultural deposits. Such use is a reflection of what was going on in archaeology generally during the first half of the twentieth century, but also is a fair caricature of what has happened in the second half of the century.

As we have seen, Ford (1938a:262) sought to identify "distinct time horizons" represented by decoration complexes, suggesting that these could serve as "stylistic time horizons" over larger areas. A. L. Kroeber (1944:104) indicated it had become common practice in South America to think of a *horizon style* as a unit "showing definably distinct features, some of which extend over a large area, so that its relations with other, more local styles serve to place these in relative time, according as the relations are of priority, consociation, or subsequence." Gordon Willey (1945:49–50) suggested two criteria were fundamental to a horizon style:

> first, that there shall be resemblance among the style groups so classed; second, that there be uniformity in relative position in the sequence on the part of the style as it occurs from region to region. Resemblances among component regional styles of a horizon style are established on the basis of very definite sets of features.

Willey (1945:53) indicated that horizons and horizon styles were "an integration of artistic elements which has been widely diffused at a given time period" and could exist at the scale of an attribute, a discrete object (type), or a set of discrete objects (multiple associated types). Such units could be used as "the horizontal stringers by which the upright columns of specialized regional development are tied together in the time chart" (Willey, 1945:53, 55). Thus the value of horizons and horizon styles resided in their indicating temporal similarity: "the presence of any style in two or more places has a high synchronic value" (Willey, 1951:111).

Willey (1945:53) defined a cultural tradition as comprising "a line, or a number of lines, of pottery development through time within the confines of a certain technique or decorative constant." Traditions demonstrated "the staying power [read *persistence*] of certain regional-cultural ideas" (Willey, 1945:56). As explained by Irving Rouse (1953:70), horizons allow "synchronizing periods over a large region," whereas traditions imply the survival of culture traits "from period to period within a single geographic area." Horizons and traditions are units meant to reveal relations among different cultures or cultural phases (Phillips and Willey, 1953; Willey and Phillips, 1958). Such integration of phases or similar units in time and space is justified in commonsense terms as exemplifying the diffusion (horizons) or persistence (traditions) of traits. Why culture traits do or do not diffuse or persist was opaque in the formulations of Phillips and Willey (1953) (Willey and Phillips, 1958).

Rouse (1955) sought ultimately to determine the phylogenetic relations among cultural phases. To argue that contemporaneous phases were genetically related "because they share a given horizon style ... is on the genetic ... level of interpretation, for it requires an assumption that the style has diffused from one phase to the others with little or no time lag" (Rouse, 1955:718). To trace the evolutionary—"genetic"—relations among phases, one first had to establish the time–space distributions of phases and their formal similarities (Rouse, 1955:719) in order to demonstrate that they had been in "contact." Second, one needed to distinguish between analogous and homologous similarity. Efforts to accomplish this step, however, were thwarted not because of any weakness in Rouse's formulation but because culture historical interests had three decades earlier become centered around stylistic phenomena that allowed one to measure time. There was no effort to build a theoretical model of phylogeneticlike culture history and no effort to develop analytical methods for distinguishing between analogous and homologous similarity. Because styles rendered as index fossils performed so well at working out chronological relations and correlating archaeological materials, culture historians assumed that they could be used for other analytical purposes as well. In lieu of theory, Willey's (1953:363) axiom that "typological similarity [denotes] cultural relatedness" was the best that could be offered.

Early on, Willey (1945:50) cautioned that "Horizon-style resemblance, although specific, does not indicate cultural identity of the groups who participated in the style." This, of course, is precisely Niles Eldredge and Stephen Jay Gould's (1977) point that we cited with respect to biostratigraphy in the beginning of this chapter. Formal similarity of fossils or artifacts does denote some equivalency, but the precise kind of equivalency thought to be represented is an inference. In the case of marker types of artifacts and horizon styles, temporal equivalency is not an unreasonable inference *if* spatial propinquity is indicated *and* either "the same types are found associated with one another repeatedly in different cultural units" or "similar features of design exist in the same complex patterned relationships in different local units" (Patterson, 1963:391). In other words, the more complex the similarities, at whatever scale, the more likely the compared phenomena are equivalent in time. And, of course, the briefer the duration is of the index fossils, the potentially closer in time the compared phenomena are (Patterson, 1963).

An inference of cultural equivalency is another thing altogether; as should be clear from our discussions in earlier chapters, not only is historical continuity demanded but so is heritable continuity. The latter requirement has seldom been acknowledged. For example, Southwesternist Emil Haury (1936b:80) described the general process of constructing spatial–temporal phases that supposedly linked culturally related sets of artifacts: first, establish pottery types; second, determine the stratigraphic positions and relations of types "to provide the primary diagnostic of phases"; and third, examine other categories of artifacts—architecture, figurines, stone tools—"to round out and complete the list of phase components"

(Haury, 1937b:19–20). Thus cultural phases are to be constructed on the basis of types that had limited time–space distributions. Some types might then become index fossils that designate phases, and in so far as such units are conceived of as cultural units in some ethnographic sense, the discovery of a particular index fossil in a new area denotes the entire ethnographic unit. This was not the only problem, as the use of strata as collection units and the use of index fossils or marker types disallowed the detection of continuous change at the level of culture units.

At the level of a cultural unit, there was no change but only transformations from one phase to the next. That is, heritable continuity was not documented in the sense that it is in seriation, but only as greater or lesser degrees of formal difference. The occasional detection of such things as early, middle, and late manifestations of a phase was the result of finding pottery with poorly developed attributes of a particular phase, other pottery with well-developed attributes of that particular phase, and still other pottery that contained a mixture of the well-developed attributes with attributes of the succeeding phase. Haury (1937a) wrote that such transitional specimens are made prominent as tangible evidence of continuity in the development of pottery and demonstrated a genetic relation between different pottery types. Although such transitional specimens demonstrate the gradual, continuous evolution of pottery, they were either ignored or cut up into stratigraphically warranted index fossils that denoted chunks of the temporal–cultural continuum termed phases, which in turn were treated as real culture units.

7 Final Thoughts on Archaeological Time

A Clash of Two Metaphysics

The study of time is an interesting topic from the standpoint not only of how archaeologists measure it but also of how perspectives on its measurement have changed throughout the twentieth century. The study of time becomes even more interesting when we examine it in the broader arena of how archaeologists come to know what they think they know about the archaeological record. This to us is one of the most fascinating, as well as one of the most underexamined, aspects of the discipline, and we can use the study of time as a vehicle through which to explore it.

Archaeology is not so different from many other disciplines in that much of what we normally assume about the way things were in the past comes from years' worth of accumulated knowledge, coupled with a healthy dose of common sense resulting from our everyday experiences with the natural world. Without thinking about it, we assume that things had to be a certain way at a given point in the past because there is general consensus on the matter. In other words, common sense and received wisdom come to dictate what is true and what is not true. There is nothing wrong with common sense; it is one of several ways in which we attempt to impose order on a complex and irregular world. This strategy works well in archaeology as long as there is unanimity of opinion as to what happened in the past, but what do we do in cases where your version of common sense tells you one thing and ours tells us something else? Do we turn the matter over to a jury and let it vote on the best scenario, being swayed, perhaps, by which of us has the most elaborate, audience-friendly, or interesting scenario? We could do this easily enough, and in fact it is done routinely in archaeology. It happens so frequently and usually so quietly that we never really pay much attention to it. We reconstruct the past by interpreting the archaeological record and then offer those interpretations up for approval by our peers. Some of the interpretations probably are accurate portrayals of the past, others might be pretty close, and still others probably are far off the mark. The problem is, how do we know which is which? The answer is, we do not. Common sense might suggest, perhaps strongly, which interpretation is the most nearly accurate one, but there still is no guarantee that it is accurate.

Science does not operate on opinion and voting. It is a special kind of sense-making system that "is distinguished by the use of theory to explain phenomena and that employs an empirical standard of truth as the ultimate arbiter of the correctness of its conclusions" (Dunnell, 1988:9). Notice that science uses theory to explain why things are the way they are, and it employs empirical measures to determine how well the conclusions stack up against what the theory tells us the conclusions should be. These empirical measures are nothing more than a coherent body of methods and techniques that are useful for determining the validity of logical implications derived directly from the theory. It is this strict reliance on the construction of propositions that can be proved false, precisely because of their empirical implications, that sets science apart from other sense-making systems.

Throughout its history, Americanist archaeology has walked a fine line between science and common sense. If we accept the above depiction of science—that it uses theory to explain natural phenomena and employs empirical standards to measure the fit between data and conclusions—then certainly much of what has been passed off in archaeology as being scientific does not and cannot fit the bill. Fieldwork and analysis have not been structured in such a way that conclusions can be falsified, instead being geared toward confirming conclusions arrived at by whatever means. Under such a modus operandi, arguments are constructed to accommodate available data. In some instances new data cause the reformulation of conclusions, but in essence what is being done is to broaden the conclusions to the point where they simultaneously "explain" everything, while actually explaining nothing. If cause is invoked, it is rendered in terms of old anthropological standbys such as acculturation, diffusion, and the like, which are, in fact, what are in need of explanation. In such cases cause is internal to the phenomena being explained.

Science locates cause in theory, making it something that is attributed rather than observed (Dunnell, 1988). Cause is external to rather than lodged in the phenomena being explained. Science operates under the assumption that, to use evolutionary biologist Richard Lewontin's (1974:8) words, "we cannot go out and describe the world in any old way we please and then sit back and demand that an explanatory and predictive theory be built on that description." Further, Lewontin (1974:8) points out that "it is not always appreciated that the problem of theory building is a constant interaction between constructing laws and finding an appropriate set of descriptive state variables (units) such that laws can be constructed." The operative words here are *appropriate units*; science fails completely as an exercise when the wrong kinds of units are used to perform analytical work. This is why we made explicit reference in Chapter 2 to the difference between ideational and empirical units. Philip Phillips, James A. Ford, and James B. Griffin understood the importance of keeping those units separate when they pointed out the difference between calling a sherd Baytown Plain and saying that "This sherd sufficiently resembles material which *for the time being* we have elected to call Baytown Plain" (Phillips *et al.*, 1951:66).

Americanist archaeologists from the start had a desire to be scientific, and it would be the height of arrogance to suggest that because they did not debate the merits of such things as predictive theory, deductive reasoning, or empirical standards that all of their efforts were unscientific. Similarly, just because archaeologists working since 1962 have debated such things does not necessarily make all that they have done scientific. Any legitimate attempt to categorize archaeological analysis requires an objective means of separating the scientific, that is, the explanatory, component from what might be labeled the interpretive component. No better place exists to begin such an examination than with time and how it has been measured. We say this because it is there where we see the conflict of two metaphysics that met head on in the late teens and set Americanist archaeology off on a trajectory much different from one that might have been predicted a few years previously. Those two metaphysics, or ontological positions, are essentialism and materialism, and the interplay between them has much relevance to modern archaeology. We bypass discussion of the philosophical underpinnings of these views of reality (see papers in O'Brien, 1996b) and focus instead on what each brings to the study of time. What we will find is that archaeologists have consistently flip-flopped in how they view time and how and why they segment it the way they do. All archaeologists would agree, and have for eight decades, that time is a continuum and that artifact types, if properly constructed, can measure the passage of time. What they have consistently ignored is that the units they employ to measure time are not real. But once those units were thought of as real, it was impossible to staunch the flow of interpretations that were spun about the archaeological record.

MEASURING TIME CONTINUOUSLY

Prior to 1910, it generally was accepted that superposed strata documented the passage of time. Attempts to use this fact during the first 10 years of the twentieth century to document and study culture change met with virtually no approval from the discipline at large because change was cast in essentialist terms, and thus only major differences in culture traits, generally construed as things rather than as kinds of things, were considered significant. Such differences, however, were seldom found. Simultaneously, attempts to standardize the description of artifacts grew in number until, by the middle of the second decade of the twentieth century, so much variation was recognized that it was suspected that some of it denoted temporal difference. By 1915, a handful of archaeologists working in New Mexico realized that artifact types could be used not simply to document the passage of time but also to measure the passage of time. They did not explicitly deny that types might measure something else as well, a holdover from the nineteenth-century view that ethnicity might be measured, but they made it clear that artifact types were archaeological constructs formulated for the purpose of telling time. Nels Nelson, for example, indicated that he could easily have constructed pottery types other than the ones he did based on his work in the Galisteo Basin, and it is obvious from examining the work that A. L. Kroeber conducted around Zuñi Pueblo that his types were artificial, the result of constructing and reconstructing types until he arrived at units that measured the passage of time in a comprehensible way. The nature of this sorting and resorting makes it clear that early culture historians were not equating types with real, empirical, units that came from the minds of prehistoric peoples. Rather, they were ideational units that were useful for partitioning time into analyzable segments.

By studying changes in the frequencies of artifact types, or variants of a culture trait—pottery, for example—time could be measured as a continuous variable. This was an entirely new way to view the passage of time archaeologically. It was not wholly new, however, in anthropology. Kroeber (1909) earlier had characterized perceived fluctuations in frequencies of trait variants as simply "passing changes of fashion" when he denounced Max Uhle's (1907) discussion of culture change in the San Francisco Bay area, and Clark Wissler (1916b:195–196) described them as "stylistic pulsations" when referring to Nelson's (1916) analysis of pottery from New Mexico. It was Nelson, though, who formalized this notion for archaeologists when he wrote that "normal frequency curves [of pottery styles indicate that types] came slowly into vogue, attained a maximum and began a gradual decline" (Nelson, 1916:167). This axiom—the *popularity principle*—was nothing more than a commonsense explanation for perceived phenomena, but it came to serve as the central tenet of a scientific archaeology.

For his part, Kroeber simultaneously and independently of any European influence invented the technique known as *frequency seriation*, concluding that the temporal implications of the arrangement of his surface-collected assemblages having different frequencies of pottery types could only be confirmed by excavation (Lyman and O'Brien, 1999). A. V. Kidder, on the other hand, copied the European technique of phyletic seriation to arrange his pottery in a temporal sequence. He, too, sought to confirm that the arrangement was chronological via stratigraphic excavation. Kidder mimicked Nelson's technique of examining fluctuating frequencies of pottery by level, later referred to variously as *ceramic*, or *percentage*, *stratigraphy* (Ford, 1962; Willey, 1938, 1939), but he used natural rather than arbitrary vertical units and relative rather than absolute frequencies of pottery types. Leslie Spier, too, seriated surface-collected assemblages based on the relative frequencies of pottery types, and he excavated in arbitrary metric levels and used percentage stratigraphy to test the chronological significance of his and Kroeber's seriations (Spier, 1917a,b).

The techniques used by Nelson, Kidder, Kroeber, Spier, and others to establish a chronology formed the basis of what later came to be known as the *direct historical approach* (Steward, 1942). Simply put, to anchor a relative chronology of pottery types, one began with the most recent, or historically known, end of the chronological continuum and then simply worked backward in time. This approach gained popularity across North America and was used with great success by William Duncan Strong (1935, 1940), Waldo Wedel (1938, 1940), and others. The development of dendrochronological methods in the Southwest helped to make relative chronologies developed there absolute and eventually led to the creation of pottery types that denoted smaller and smaller chunks of the temporal continuum.

The effect of Nelson's, Kroeber's, Kidder's, and Spier's efforts on Americanist archaeology was immediate. No longer was the apparent absence of an extensive time depth to American prehistory a problem in attempts to write the history of ethnographically documented cultures; no longer did one have to search for differences in superposed trait lists; and no longer did one have to search for elusive stratified deposits. Artifacts from any archaeological deposit of sufficient thickness could be chronologically ordered, as Spier showed, if those artifacts were categorized in particular ways. A chronological ordering based on seriation, either of Kroeber's frequency sort or of Kidder's phyletic sort, was testable when stratified deposits were available or if one used dendrochronology. Thus *only* the kinds of types that allowed the measurement of time—those that passed the historical significance test (Krieger, 1944)—were desired because they allowed one to produce a testable product: a chronological sequence. The subsequent focus of classification efforts around this singular goal was due to the fact that "for the first time it was possible to do archaeology and be wrong! This jerked archaeology out of the business of speculative natural history and placed it firmly in the realm of science" (Dunnell, 1986b:29).

The so-called stratigraphic revolution was less a revolution in how sites were excavated and artifacts collected and much more a revolution in how artifacts were classified. A good classification was one that produced types that allowed time to be measured. Stratigraphy played an important role, but it was a confirmational role rather than a role of creation or discovery. A chronological order was constructed using seriation, and stratigraphic observation was used to test the ordering to ensure that it was measuring the passage of time correctly. Inferences regarding the passage of time, which were derived using one method, could actually be tested and either empirically confirmed or shown to be wrong through the use of a separate method. Stratigraphy, or more correctly, the principle of superposition, was the ultimate arbiter in questions of chronology. With confirmation of Kidder's, Nelson's, and Kroeber's pottery sequences in hand, Wissler (1917b) proclaimed that a "new archaeology" had emerged.

MEASURING TIME DISCONTINUOUSLY

If we stopped the story at this point, one might speculate that the future of Americanist archaeology lay in the direction of improving the chronological methods that had been developed in New Mexico. Although that speculation would have some veracity, it would overlook the more significant direction the discipline took. Perhaps the point is arguable, but in our opinion this is the critical juncture where archaeology gave back much of the ground it had recently acquired. It is ironic that Wissler heralded the chronological triumphs in the Southwest as the "new archaeology" because 50 years later the same term would be used to usher in a period when Americanist archaeology pledged its allegiance to anthropology.

By the end of the second decade of the twentieth century, archaeologists became discontented with simply using artifact types to mark the passage of time. Americanist archaeologists are, after all, trained as anthropologists, and maybe it was only natural that they would begin to use types the same way that their predecessors in the nineteenth century had: to identify specific groups of people. But because of the recognition that there was considerable time depth to the archaeological record, archaeologists were now using artifact types to track prehistoric groups across space and time. Some archaeologists, mostly those affiliated with the American Museum, examined the relative frequencies of artifact types in order to measure time and to discuss cultural development as a continuum, but most other archaeologists working after about 1920 did not. The majority simply looked for superposed materials and then labeled aggregates of materials "cultures" or the like. This was precisely the conception Franz Boas had when he directed Manuel Gamio to excavate stratigraphically in the Valley of Mexico, a patently essentialist view.

Cultures, so the thinking went, were clearly evident in the ethnological record, and the traits of those cultures displayed unique and continuous distributions that could be accounted for with such notions as the culture area concept (Wissler, 1916a) and the age–area concept (Wissler, 1919). Wissler (1919) showed that the culture area and age–area notions were fully applicable to, confirmed by, and accounted for the archaeological record. By 1920, the critical issues were how artifacts were to be sorted and studied so that such explanatory tools could be called upon and how cultures as larger-scale units were to be recognized and/or constructed. Stratigraphic excavation no longer was used in a confirmational role; it now assumed the role of a method for discovering chronologies, not simply chronologies of pottery types but chronologies of cultures. The shift in the role, not in the perception, of vertically superposed units ultimately led to various conceptions of mixed and reversed stratigraphy when what were thought to be the diagnostic remains of distinct cultures were found within the same vertical excavation unit or when a suspected pottery sequence was found to be out of order.

Thus, the discipline appears to have accepted Wissler's impression that the presence of empirically discrete and distinct depositional units (strata) denoted discontinuous occupation by multiple successive cultures. The belief that vertically stacked units could be used to denote discrete "occupations" reinforced otherwise nebulous notions of a sequence of visibly distinct "cultures," phenomena that were readily perceived in the ethnographic and spatial records. Even such luminaries as Kidder, who had demonstrated through phyletic seriation that it was possible to measure time continuously, began to discuss such things as "ceramic periods," though he noted simultaneously that the particular pottery types denoting the periods were nothing more than "useful cross-sections of a constantly changing cultural trait [that had undergone] a slow, usually subtle, but never-ceasing metamorphosis" (Kidder, 1936a:xx–xxi). Kidder's comments mirrored those of Nelson (1932:105), who made it clear that the history of a culture was conceived of as a stream, the flow rate of which could be measured by observing "a few cross-sections of the flow taken at strategic points." Those cross sections were conceived of as discrete occupations thought to be manifest within distinct depositional units usually termed strata. Therein lay the problem: conflation of opposing metaphysics.

An example of the problem built into archaeology from the time when superposition became creational rather than confirmational is found in Gordon Willey's (1939) efforts to understand the cultural chronology of Georgia and Florida. Willey found that in some cases stratigraphic separation of what he viewed as cultural layers was ambiguous, a result of what he termed stratigraphic "disturbance" and "mixing," and he later turned to index fossils to make the separation. His later wording sounds a lot like Kidder's:

> Any pottery type is based on a number of stylistic features found in combination, but changes occur through time, and transitions are often so gradual as to prevent sharp distinctions. However, the periods into which the pottery types have been grouped are each based on one or more "key" or "marker" types, which have been found to be sufficiently restricted in range and distinctive in appearance to allow their occurrence to be quite positively determined. It is also hoped that the "periods" will prove to represent distinct cultures when the bare skeleton of ceramic chronology has been given flesh and body in the form of a full and "functional" culture description. (Willey and Woodbury, 1942:236)

Willey's problem was the same as Kidder's: cultural development conceived in continuous terms but perceived, or at least thought only to be measurable, in discontinuous, "real" terms. Transitional specimens were the obvious results of this perception combined with the failure to distinguish between empirical and theoretical units.

The quotation from Willey and Woodbury exemplifies the conflation of a conceptual model of gradual, continuous cultural change—a flowing stream—with the ethnologically informed perception of cultures as real, discrete units. Index fossils, or marker types, could be used to measure the passage of time, but it was also possible to use them as icons signifying an ethnographic unit. Even though such a correlation was untestable, it served as an ad hoc, commonsense warrant for dividing the cultural continuum into chunks referred to as cultures. Synthesis of an area's prehistory, whether with the Midwestern taxonomic method

(McKern, 1934, 1937, 1939) or the later system of Willey and Phillips (1958), focused on precisely this kind of unit. Thus archaeologists viewed types not only as analytical units allowing the measurement of time but also as accurate reflections of distinct ethnographic units.

By the 1940s, artifact types were dual purpose units. First, they were constructed to measure time, and their usefulness as such could be ascertained with the historical significance test. If artifact types represented in collections were seriated or plotted against their stratigraphic provenience, they should display continuity and unimodal frequency distributions. Such indicated that the types reflected the waxing and waning of a culture trait variant's popularity through, and thus served well as a measure of, time. Second, if types in two different time and/or space positions were similar, then they must represent continuity of the ideas that underlay them as well, such as if two cultural streams of flowing ideas had intersected. In other words, similar types were interpreted as representing some sort of common ancestry. It was an interpretation or inference because of the lack of an explicit theory of cultural evolution written in Darwinian terms such as heritable continuity. Because of this lack, interpretations of common ancestry or that similarity was homologous were debatable and incapable of testing; analogues and homologues could not be distinguished, nor could synapomorphic and symplesiomorphic traits. Instead, historical continuity rendered as typological similarity was interpreted as heritable continuity.

Early attempts to derive evolutionary histories and construct artifact lineages from the archaeological record (e.g., Kidder, 1917; Kidder and Kidder, 1917) for the most part were unnoticed by the 1940s, probably because the use of evolutionary wording such as "one type had descended from another" was largely metaphorical. When the wording became more literal, as it did under the guidance of Harold S. Colton (e.g., 1939; Colton and Hargrave, 1937), a biologist by professional training, and Harold S. Gladwin (e.g., Gladwin and Gladwin, 1934), the reaction was swift and sure (e.g., Reiter, 1938; Steward, 1941). The fact that one of the proponents quickly abandoned many of the connotations of such wording (Gladwin, 1936) signified the beginning of the end for biologically based evolutionary wording in archaeology. J. O. Brew's (1946) remarks that biological evolution would not resolve the dilemma put an end to such efforts (Lyman and O'Brien, 1997).

It was clear that artifact types, if constructed in particular ways, could be used quite successfully to measure the flight of time's arrow. One way to explain this utility was to embue them with "cultural" meaning; artifact types changed from one state to another as culture types changed from one state to another. A change in the former signified, then, a change in the latter. Conceiving of types as discontinuous parts of the temporal—and the cultural—continuum directly contradicted the model depicted in Fig. 3.7. By deriving types extensionally from stratigraphically bounded units, using superposition and index fossils to measure time, and largely discarding the seriation method (the effects of the first two being exacerbated by dendrochronology and radiocarbon dating), the contradiction was "empirically" reinforced to such an extent that by the 1950s the preferred model was no longer one of continuous culture change.

Recall Kidder's (1936a:xx) early suggestion that "recognizable nodes of individuality" in pottery types represented ceramic and cultural periods. Phillips and Willey (1953:622) suggested that a culture could, "under certain circumstances, be remarkably stable"; that is, no or minimal change would occur. A few years later, this notion received support from participants in the Society for American Archaeology's 1955 Seminars on archaeological method and theory. They stated that "shifts from one culture type to another are comparatively rapid" and that were culture change to be modeled as a graph of cultural variation against time, the graph would show a curve "characterized by a series of sharply rising escarpments connected by slightly sloping plateaus" (Thompson, 1956:37). Transitional culture types would be rare in the archaeological record if they had existed at all. Finally, Albert C. Spaulding (in Willey and Phillips, 1958:15–16) argued that "widely accepted cultural theory indicates that the normal pattern [of culture change] is one of relative stability, then rapid growth through the introduction of a critical new [culture trait] followed very quickly by a number of other new [traits], then a period of relative stability, and so on." Culture change was no longer conceived of as continuous but rather as discontinuous. The techniques archaeologists of the mid-twentieth century used to measure relative time led them into a trap from which the discipline has not yet escaped, despite explicit recognition of the trap for over five decades (e.g., Brew, 1946; Dunnell, 1995; Plog, 1974). The trap is the essentialist–materialist paradox.

CONCLUDING REMARKS

Are the efforts of Nelson, Kroeber, Kidder, Spier, Ford, Willey, and others unworthy of our attention because those archaeologists became ensnared in the essentialist–materialist paradox? Not at all. Understanding the issues raised in the preceding pages of this chapter allows us to see not only why and how Americanist archaeology took the form it did but also the ingenuity that our predecessors had relative to developing methods for marking the passage of time. The fact that the methods were corrupted through unwarranted speculation in no way implies that they are not valid chronological methods. And neither does it take away from the fact that what those archaeologists did was novel and in many ways ingenious. The ingenuity was not that they began excavating stratigraphically and paying attention to superposition; Americanists had long been doing so. There was no "stratigraphic revolution" in Americanist archaeology. Rather, there was a revolution in how time was perceived and measured.

Nor were there "revolutionaries" in the sense that historians of archaeology use the term. When Gamio, Nelson, Kidder, and Spier excavated stratigraphically, each addressed a specific and local chronological problem that grew from their extensive knowledge of the areas in which they were working. They perceived what they believed were chronological differences in the materials they were studying and wanted to confirm those perceptions empirically. They realized that one had to look at variation among artifacts, not just the presence or absence of cultural traits such as pottery, to analytically detect the passage of time. To measure and document that that variation signified the flight of time's arrow, they constructed historical types. This was what was significant about their work.

One could, we suppose, adopt the attitude that none of this matters since radiometric dating has alleviated our chronological problems, thus rendering any consideration of seriation, stratigraphy, and index fossils moot. But we would argue that knowledge of how archaeological methods of relative dating work is crucial to successful archaeological research. This is so for two reasons, one of which we noted in Chapter 1 and the other of which we have made reference to throughout the other chapters. First, absolute radiometric methods are no panacea (e.g., O'Connell and Allen, 1998); one needs to evaluate and test the results obtained from the application of these methods, and relative dating methods provide one source of test implications. As well, radiometric methods may not always be applicable given the vagaries of the processes that formed and continue to form the archaeological record. Second, the relative dating methods we have discussed here, particularly seriation and cross dating, depend on a theory of culture change that is well, if incompletely, characterized as Darwinian evolution. This is not surprising, given that the methods we have discussed were developed within a research framework that had as its goals the documentation and explanation of change and development within and between cultural lineages (Lyman and O'Brien, 1997, 1998). Discarding these relative dating methods in favor of more recently invented radiometric techniques potentially results in part in concomitant disposal of the research framework. We firmly believe that the only way to escape the essentialist–materialist paradox faced by modern archaeology is to retain and to understand and *supplement* with radiometric dating techniques the relative dating methods we have discussed here.

References

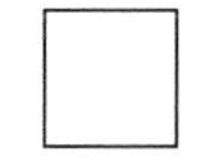

Adams, R. E.
1960 Manuel Gamio and Stratigraphic Excavation. *American Antiquity* 26:99.

Adams, W. Y.
1988 Archaeological Classification: Theory versus Practice. *Antiquity* 61:40–56.

Adams, W. Y., and Adams, E. W.
1991 *Archaeological Typology and Practical Reality: A Dialectical Approach to Artifact Classification and Sorting*. Cambridge University Press, Cambridge, England.

Ahler, S. R.
1993 Stratigraphy and Radiocarbon Chronology of Modoc Rock Shelter, Illinois. *American Antiquity* 58:462–489.

Aitken, M. J.
1985 *Thermoluminescence Dating*. Academic Press, Orlando, FL.
1990 *Science-Based Dating in Archaeology*. Longman, London.

Anonymous
1915 Museum Notes. *American Museum Journal* 15:30.

Ascher, R.
1963 Review of "A Quantitative Method for Deriving Cultural Chronology" by J. A. Ford. *American Antiquity* 28:570–571.

Bailey, G. N.
1983 Concepts of Time in Quaternary Prehistory. *Annual Review of Anthropology* 12:165–192.

Baillie, M. G. L.
1982 *Tree-Ring Dating and Archaeology*. University of Chicago Press, Chicago.

Bell, R. E.
1958 Guide to the Identification of Certain American Indian Projectile Points. *Oklahoma Anthropological Society, Special Bulletin* No. 1.
1960 Guide to the Identification of Certain American Indian Projectile Points. *Oklahoma Anthropological Society, Special Bulletin* No. 2.

Bell, R. E., and Hall, R. S.
1953 Selected Projectile Point Types of the United States. *Oklahoma Anthropological Society Bulletin* 1:1–16.

Berger, R., and Suess, H. E. (Editors)
1979 *Radiocarbon Dating*. University of California Press, Berkeley.

Biers, W. R.
1992 *Art, Artefacts, and Chronology in Classical Archaeology*. Routledge, New York.

Binford, L. R.
1962 A New Method of Calculating Dates from Kaolin Pipe Stem Samples. *Southeastern Archaeological Conference Newsletter* 9(1):19–21.
1972 The "Binford" Pipe Stem Formula: A Return from the Grave. *Conference on Historic Site Archaeology Papers, 1971* 6:230–253.

Blackwell, B. A., and Schwarcz, H. P.
1993 Archaeochronology and Scale. In Effects of Scale on Archaeological and Geoscientific Perspectives, edited by J. K. Stein and A. R. Lense, pp. 39–58. *Geological Society of America, Special Paper* 283. Boulder, CO.
Boas, F.
1900 Conclusion. In The Jesup North Pacific Expedition, edited by F. Boas, pp. 387–390. *American Museum of Natural History, Memoirs* 2(4).
1902 Some Problems in North American Archaeology. *American Journal of Archaeology* 6:1–6.
Bock, W. J.
1977 Foundations and Methods of Evolutionary Classification. In *Major Patterns in Vertebrate Evolution*, edited by M. K. Hecht, P. C. Goody, and B. M. Hecht, pp. 851–895. Plenum Press, New York.
Bowman, S.
1990 *Radiocarbon Dating*. University of California Press, Berkeley.
Bradley, B. A.
1991 Flaked Stone Technology in the Northern High Plains. In *Prehistoric Hunters of the High Plains*, 2nd edition, edited by G. C. Frison, pp. 369–395. Academic Press, San Diego, CA.
1997 Sloan Site Biface and Projectile Point Technology. In *Sloan: A Paleoindian Dalton Cemetery in Arkansas*, edited by D. F. Morse, pp. 53–57. Smithsonian Institution Press, Washington, DC.
Brainerd, G. W.
1951a The Place of Chronological Ordering in Archaeological Analysis. *American Antiquity* 16:301–313.
1951b The Use of Mathematical Formulations in Archaeological Analysis. In Essays on Archaeological Methods, edited by J. B. Griffin, pp. 117–127. *University of Michigan, Museum of Anthropology, Anthropological Papers* No. 8.
Braun, D. P.
1985 Absolute Seriation: A Time-Series Approach. In *For Concordance in Archaeological Analysis*, edited by C. Carr, pp. 509–539. Westport, Boulder, CO.
1987 Coevolution of Sedentism, Pottery Technology, and Horticulture in the Central Midwest, 200 BC–AD 600. In Emergent Horticultural Economies of the Eastern Woodlands, edited by W. F. Keegan, pp. 153–181. *Southern Illinois University, Center for Archaeological Investigations, Occasional Paper* No. 7.
Breternitz, D. A.
1966 An Appraisal of Tree-Ring Dated Pottery in the Southwest. *University of Arizona Anthropological Papers* No. 10.
Brew, J. O.
1946 The Archaeology of Alkali Ridge, Southeastern Utah. *Peabody Museum of Archaeology and Ethnology, Papers* 21.
Browman, D. L., and Givens, D. R.
1996 Stratigraphic Excavation: The First "New Archaeology." *American Anthropologist* 98: 80–95.
Coe, J. L.
1964 The Formative Cultures of the Carolina Piedmont. *American Philosophical Society, Transactions* 54(5).
Collins, H. B., Jr.
1927 Archaeological Work in Louisiana and Mississippi. *Explorations and Field-Work of the Smithsonian Institution in 1931*, pp. 200–207. Smithsonian Institution, Washington, DC.
1932 Archaeology of Mississippi. In *Conference on Southern Pre-History*, pp. 37–42. National Research Council, Washington, DC.
Colman, S. M., Pierce, K. L., and Birkeland, P. W.
1987 Suggested Terminology for Quaternary Dating Methods. *Quaternary Research* 28:314–319.

Colton, H. S.
1932 A Survey of Prehistoric Sites in the Region of Flagstaff, Arizona. *Bureau of American Ethnology, Bulletin* 104.
1939 Prehistoric Culture Units and Their Relationships in Northern Arizona. *Museum of Northern Arizona, Bulletin* 17.
Colton, H. S., and Hargrave, L. L.
1937 Handbook of Northern Arizona Pottery Wares. *Museum of Northern Arizona, Bulletin* 11.
Cowgill, G. L.
1963 Review of "A Quantitative Method for Deriving Cultural Chronology" by J. A. Ford. *American Anthropologist* 65:696–699.
1968 Review of "Computer Analysis of Chronological Seriation" by F. Hole and M. Shaw. *American Antiquity* 33:517–519.
1972 Models, Methods and Techniques for Seriation. In *Models in Archaeology*, edited by D. L. Clarke, pp. 381–424. Methuen, London.
Cracraft, J.
1981 Pattern and Process in Paleobiology: The Role of Cladistic Analysis in Systematic Paleontology. *Paleobiology* 7:456–468.
Cressman, L. S.
1951 Western Prehistory in the Light of Carbon 14 Dating. *Southwestern Journal of Anthropology* 7:289–313.
Dall, W. H.
1877 On Succession in the Shell-Heaps of the Aleutian Islands. *Contributions to North American Ethnology* 1:41–91.
Darwin, C.
1859 *On the Origin of Species*. Murray, London.
David, N.
1972 On the Life Span of Pottery, Type Frequencies, and Archaeological Inference. *American Antiquity* 37:141–142.
Davis, D. D.
1981 Ceramic Classification and Temporal Discrimination: A Consideration of Later Prehistoric Stylistic Change in the Mississippi River Delta. *Mid-Continental Journal of Archaeology* 6:55–89.
Dean, J. S.
1978 Independent Dating in Archaeological Analysis. In *Advances in Archaeological Method and Theory*, Vol. 1, edited by M. B. Schiffer, pp. 223–255. Academic Press, New York.
1993 Geoarchaeological Perspectives on the Past: Chronological Considerations. In Effects of Scale on Archaeological and Geoscientific Perspectives, edited by J. K. Stein and A. R. Lense, pp. 59–65. *Geological Society of America, Special Paper* 283. Boulder, CO.
Deetz, J.
1967 *Invitation to Archaeology*. Natural History Press, Garden City, NY.
Deetz, J., and Dethlefsen, E.
1965 The Doppler Effect and Archaeology: A Consideration of the Spatial Aspects of Seriation. *Southwestern Journal of Anthropology* 21:196–206.
1971 Some Social Aspects of New England Colonial Mortuary Art. In Approaches to the Social Dimensions of Mortuary Practices, edited by J. A. Brown, pp. 30–38. *Society for American Archaeology, Memoir* No. 25.
Dempsey, P., and Baumhoff, M.
1963 The Statistical Use of Artifact Distributions to Establish Chronological Sequence. *American Antiquity* 28:496–509.
Dethlefsen, E., and Deetz, J.
1966 Death's Heads, Cherubs, and Willow Trees: Experimental Archaeology in Colonial Cemeteries. *American Antiquity* 31:502–510.

Dewar, R. E.
1992 Incorporating Variation in Occupation Span into Settlement-Pattern Analysis. *American Antiquity* 56:604–620.
Douglass, A. E.
1929 The Secret of the Southwest Solved by Talkative Tree Rings. *National Geographic* 56: 736–770.
Drennan, R. D.
1976 A Refinement of Chronological Seriation Using Nonmetric Multidimensional Scaling. *American Antiquity* 41:290–302.
Dunnell, R. C.
1970 Seriation Method and Its Evaluation. *American Antiquity* 35:305–319.
1971 *Systematics in Prehistory*. Free Press, New York.
1978 Style and Function: A Fundamental Dichotomy. *American Antiquity* 43:192–202.
1981 Seriation, Groups, and Measurements. In *Manejo de Datos y Metodos Matematicos de Arqueologia*, compiled by G. L. Cowgill, R. Whallon, and B. S. Ottaway, pp. 67–90. Union Internacional de Ciencias Prehistoricas y Protohistoricas, Mexico DF.
1985 Archaeological Survey in the Lower Mississippi Alluvial Valley, 1940–1947: A Landmark Study in American Archaeology. *American Antiquity* 50:297–300.
1986a Methodological Issues in Americanist Artifact Classification. In *Advances in Archaeological Method and Theory*, Vol. 9, edited by M. B. Schiffer, pp. 149–207. Academic Press, New York.
1986b Five Decades of American Archaeology. In *American Archaeology: Past and Future*, edited by D. J. Meltzer, D. D. Fowler, and J. A. Sabloff, pp. 23–49. Smithsonian Institution Press, Washington, DC.
1988 Archaeology and Evolutionary Theory. Unpublished paper delivered at the University of Missouri—Columbia.
1989 Aspects of the Application of Evolutionary Theory in Archaeology. In *Archaeological Thought in America*, edited by C. C. Lamberg-Karlovsky, pp. 35–49. Cambridge University Press, Cambridge.
1995 What Is It That Actually Evolves? In *Evolutionary Archaeology: Methodological Issues*, edited by P. A. Teltser, pp. 33–50. University of Arizona Press, Tucson.
Dunnell, R. C., and Feathers, J. K.
1991 Late Woodland Manifestations of the Malden Plain, Southeast Missouri. In *Stability, Transformation, and Variation: The Late Woodland Southeast*, edited by M. S. Nassaney and C. R. Cobb, pp. 21–45. Plenum Press, New York.
Ehrich, R. W.
1950 Some Reflections on Archeological Interpretation. *American Anthropologist* 52:468–482.
Eighmy, J. L., and Sternberg, R. S. (Editors)
1990 *Archaeomagnetic Dating*. University of Arizona Press, Tucson.
Eldredge, N.
1982 Phenomenological Levels and Evolutionary Rates. *Systematic Zoology* 31:338–347.
Eldredge, N., and Gould, S. J.
1977 Evolutionary Models and Biostratigraphic Strategies. In *Concepts and Methods of Biostratigraphy*, edited by E. G. Kauffman and J. E. Hazel, pp. 25–40. Dowden, Hutchinson & Ross, Stroudsburg, PA.
Eldredge, N., and Novacek, M. J.
1985 Systematics and Paleobiology. *Paleobiology* 11:65–74.
Evans, J.
1850 On the Date of British Coins. *The Numismatic Chronicle and Journal of the Numismatic Society* 12(4):127–137.
Figgins, J. D.
1927 The Antiquity of Man in America. *Natural History* 27:229–239.

Fisher, D. C.
1994 Stratocladistics: Morphological and Temporal Patterns and Their Relation to Phylogenetic Process. In *Interpreting the Hierarchy of Nature*, edited by L. Grande and O. Rieppel, pp. 133–171. Academic Press, San Diego, CA.

Fleming, S.
1976 *Dating in Archaeology: A Guide to Scientific Techniques*. St. Martin's Press, New York.

Ford, J. A.
1935a An Introduction to Louisiana Archaeology. *Louisiana Conservation Review* 4(5):8–11.
1935b Outline of Louisiana and Mississippi Pottery Horizons. *Louisiana Conservation Review* 4(6):33–38.
1935c Ceramic Decoration Sequence at an Old Indian Village Site near Sicily Island, Louisiana. *Louisiana Department of Conservation, Anthropological Study* No. 1.
1936a Analysis of Indian Village Site Collections from Louisiana and Mississippi. *Louisiana Department of Conservation, Anthropological Study* No. 2.
1936b Archaeological Methods Applicable to Louisiana. *Louisiana Academy of Sciences, Proceedings* 3:102–105.
1938a A Chronological Method Applicable to the Southeast. *American Antiquity* 3:260–264.
1938b *An Examination of Some Theories and Methods of Ceramic Analysis*. MA thesis, University of Michigan, Ann Arbor.
1940 Review of *Handbook of Northern Arizona Pottery Wares* by H. S. Colton and L. L. Hargrave. *American Antiquity* 5:263–266.
1949 Cultural Dating of Prehistoric Sites in Virú Valley, Peru. *American Museum of Natural History, Anthropological Papers* 43(1):29–89.
1951 Greenhouse: A Troyville–Coles Creek Period Site in Avoyelles Parish, Louisiana. *American Museum of Natural History, Anthropological Papers* 44(1):1–132.
1952 Measurements of Some Prehistoric Design Developments in the Southeastern States. *American Museum of Natural History, Anthropological Papers* 43(3):313–384.
1954a Comment on A. C. Spaulding's "Statistical Techniques for the Discovery of Artifact Types." *American Antiquity* 19:390–391.
1954b On the Concept of Types: The Type Concept Revisited. *American Anthropologist* 56:42–53.
1962 A Quantitative Method for Deriving Cultural Chronology. *Pan American Union, Technical Bulletin* No. 1.

Ford, J. A., and Griffin, J. B.
1938 Report of the Conference on Southeastern Pottery Typology. Mimeographed. [Reprinted in *Newsletter of the Southeastern Archaeological Conference* 7(1):10–22, and in Lyman *et al.*, 1997a]

Fortey, R. A.
1985 Gradualism and Punctuated Equilibria as Competing and Complementary Theories. In Evolutionary Case Histories from the Fossil Record, edited by J. C. W. Cope and P. W. Skelton, pp. 17–28. *Palaeontological Association, Special Papers in Palaeontology* No. 33.

Fowke, G.
1922 Archeological Investigations. *Bureau of American Ethnology, Bulletin* 76.

Futuyma, D. J.
1986 *Evolutionary Biology*, 2nd edition. Sinauer, Sunderland, MA.

Gaffney, E. S., Dingus, L., and Smith, M. K.
1995 Why Cladistics? *Natural History* 104(6):33–35.

Gamio, M.
1913 Arqueologia de Atzcapotzalco, D.F., Mexico. *Eighteenth International Congress of Americanists, Proceedings*, pp. 180–187. London.
1917 Investigaciones Arqueológicas en México. *Nineteenth International Congress of Americanists, Proceedings*, pp. 125–133. Mexico City.
1924 The Sequence of Cultures in Mexico. *American Anthropologist* 26:307–322.

Gary, M., McAfee, Jr., R., and Wolf, C. L. (Editors)
1974 *Glossary of Geology*. American Geological Institute, Washington, DC.
Gayton, A. H.
1927 The Uhle Collections from Nievería. *University of California, Publications in American Archaeology and Ethnology* 21(8):305–329.
Gayton, A. H., and Kroeber, A. L.
1927 The Uhle Pottery Collections from Nazca. *University of California, Publications in American Archaeology and Ethnology* 24(1):1–46.
Geyh, M. A., and Schleicher, H.
1990 *Absolute Age Determination: Physical and Chemical Dating Methods and Their Applications*. Springer-Verlag, Berlin.
Ghiselin, M. T.
1981 Categories, Life and Thinking. *Behavioral and Brain Sciences* 4:269–283.
Gibson, J. L.
1982 *Archeology and Ethnology on the Edges of the Atchafalaya Basin, South Central Louisiana*. Report submitted to the U.S. Army Corps of Engineers, New Orleans.
Gifford, J. C.
1960 The Type–Variety Method of Ceramic Classification as an Indicator of Cultural Phenomena. *American Antiquity* 25:341–347.
Gingerich, P. D.
1976 Cranial Anatomy and Evolution of Early Tertiary Plesiadapidae. *University of Michigan Papers in Paleontology* 15:1–140.
1979 The Stratophenetic Approach to Phylogeny Reconstruction in Vertebrate Paleontology. In *Phylogenetic Analysis and Paleontology*, edited by J. Cracraft and N. Eldredge, pp. 41–77. Columbia University Press, New York.
Gingerich, P. D., and Simons, E. L.
1977 Systematics, Phylogeny, and Evolution of Early Eocene Adapidae (Mammalia, Primates) in North America. *University of Michigan, Museum of Paleontology, Contributions* 24: 245–279.
Givens, D. R.
1992 *Alfred Vincent Kidder and the Development of Americanist Archaeology*. University of New Mexico Press, Albuquerque.
1996 History of Archaeology from 1900 to 1950: Archaeology of the Americas. In *The Oxford Companion to Archaeology*, edited by B. M. Fagan, pp. 295–296. Oxford University Press, New York.
Gladwin, H. S.
1936 Editorials: Methodology in the Southwest. *American Antiquity* 1:256–259.
Gladwin, W., and Gladwin, H. S.
1928 The Use of Potsherds in an Archaeological Survey of the Southwest. *Medallion Papers* No. 2.
1930 A Method for the Designation of Southwestern Pottery Types. *Medallion Papers* No. 7.
1934 A Method for the Designation of Cultures and Their Variations. *Medallion Papers* No. 15.
Goodyear, A. C.
1982 The Chronological Position of the Dalton Horizon in the Southeastern United States. *American Antiquity* 47:382–395.
Gould, S. J.
1987 *Time's Arrow, Time's Cycle: Myth and Metaphor in the Discovery of Geological Time*. Harvard University Press, Cambridge, MA.
Graves, M. W.
1982 Breaking Down Ceramic Variation: Testing Models of White Mountain Redware Design Style Development. *Journal of Anthropological Archaeology* 1:305–354.
1984 Temporal Variation among White Mountain Redware Design Styles. *The Kiva* 50:3–24.

Graves, M. W., and Cachola-Abad, K.
1996 Seriation as a Method of Chronologically Ordering Architectural Design Traits: An Example from Hawai'i. *Archaeology in Oceania* 31:19–32.
Grayson, D. K.
1983 *The Establishment of Human Antiquity*. Academic Press, New York.
Hancock, J. M.
1977 The Historic Development of Concepts of Biostratigraphic Correlation. In *Concepts and Methods of Biostratigraphy*, edited by E. G. Kauffman and J. E. Hazel, pp. 3–22. Dowden, Hutchinson & Ross, Stroudsburg, PA.
Hanson, L. H., Jr.
1971 Kaolin Pipe Stems—Boring in on a Fallacy. *Conference on Historic Site Archaeology Papers, 1969* 4:2–15.
1972 A Few Cents More. *Conference on Historic Site Archaeology Papers, 1971* 6:254–257.
Hargrave, L. L.
1932 Guide to Forty Pottery Types from the Hopi Country and the San Francisco Mountains, Arizona. *Museum of Northern Arizona*, Bulletin 1.
Harper, C. W., Jr.
1980 Relative Age Inference in Paleontology. *Lethaia* 13:239–248.
Harrington, J. C.
1954 Dating Stem Fragments of Seventeenth and Eighteenth Century Clay Tobacco Pipes. *Archaeological Society of Virginia, Quarterly Bulletin* 9(1):9–13.
Harrington, M. R.
1909a Ancient Shell Heaps near New York City. In The Indians of Greater New York and the Lower Hudson, edited by C. Wissler. *American Museum of Natural History, Anthropological Papers* 3:167–179.
1909b The Rock-Shelters of Armonk, New York. In The Indians of Greater New York and the Lower Hudson, edited by C. Wissler. *American Museum of Natural History, Anthropological Papers* 3:123–138.
1924 An Ancient Village Site of the Shinnecock Indians. *American Museum of Natural History, Anthropological Papers* 22(5):227–283.
Haury, E. W.
1936a Some Southwestern Pottery Types, Series IV. *Medallion Papers* No. 19.
1936b The Mogollon Culture of Southwestern New Mexico. *Medallion Papers* No. 20.
1937a Pottery Types at Snaketown. In Excavations at Snaketown: I. Material Culture, by H. S. Gladwin, E. W. Haury, E. B. Sayles, and N. Gladwin, pp. 169–229. *Medallion Papers* No. 25.
1937b Stratigraphy. In Excavations at Snaketown: I. Material Culture, by H. S. Gladwin, E. W. Haury, E. B. Sayles, and N. Gladwin, pp. 19–35. *Medallion Papers* No. 25.
Hawkes, E. W., and Linton, R.
1917 A Pre-Lenape Culture in New Jersey. *American Anthropologist* 19:487–494.
Hawley, F. M.
1937 Reversed Stratigraphy. *American Antiquity* 2:297–299.
Heizer, R. F. (Editor)
1959 *The Archaeologist at Work*. Harper, New York.
Hester, T. R., Heizer, R. F., and Graham, J. A.
1975 *Field Methods in Archaeology*, 6th edition. Mayfield, Palo Alto, CA.
Hole, F., and Heizer, R. F.
1973 *An Introduction to Prehistoric Archaeology*, 3rd edition. Holt, Rinehart and Winston, New York.
Hole, F., and Shaw, M.
1967 Computer Analysis of Chronological Seriation. *Rice University Studies* 53(3).
Holmes, W. H.
1886 Ancient Pottery of the Mississippi Valley. *Bureau of Ethnology, Annual Report* 4:361–436.

1892 Modern Quarry Refuse and the Paleolithic Theory. *Science* 20:295–297.
1893 Are There Traces of Man in the Trenton Gravels? *Journal of Geology* 1:15–37.
1897 Stone Implements of the Potomac–Chesapeake Tidewater Province. *Bureau of American Ethnology, Annual Report* 15:13–152.

Howard, E. B.
1935 Evidence of Early Man in America. *The Museum Journal* 24:53–171.

Hull, D.
1978 A Matter of Individuality. *Philosophy of Science* 45:335–360.
1980 Individuality and Selection. *Annual Review of Ecology and Systematics* 11:311–332.

Jefferson, T.
1801 *Notes on the State of Virginia*. Furman and Loudan, New York.

Johnson, L., Jr.
1972 Introduction to Imaginary Models for Archaeological Scaling and Clustering. In *Models in Archaeology*, edited by D. L. Clarke, pp. 309–379. Methuen, London.

Judd, N. M.
1929 The Present Status of Archaeology in the United States. *American Anthropologist* 31: 401–418.

Justice, N. D.
1987 *Stone Age Spear and Arrow Points of the Midcontinental and Eastern United States*. Indiana University Press, Bloomington.

Kidder, A. V.
1915 Pottery of the Pajarito Plateau and Some Adjacent Regions in New Mexico. *American Anthropological Association, Memoirs* 2:407–462.
1916 Archeological Explorations at Pecos, New Mexico. *National Academy of Sciences, Proceedings* 2:119–123.
1917 A Design-Sequence from New Mexico. *National Academy of Sciences, Proceedings* 3:369–370.
1919 Review of "An Outline for a Chronology of Zuñi Ruins; Notes on Some Little Colorado Ruins; Ruins in the White Mountains, Arizona" by L. Spier. *American Anthropologist* 21:296–301.
1924 An Introduction to the Study of Southwestern Archaeology, with a Preliminary Account of the Excavations at Pecos. *Papers of the Southwestern Expedition, Phillips Academy* No. 1. Yale University Press, New Haven, CT.
1931 The Pottery of Pecos, Vol. 1. *Papers of the Southwestern Expedition, Phillips Academy* No. 5. Yale University Press, New Haven, CT.
1936a Introduction. In The Pottery of Pecos, Vol. II, by A. V. Kidder and A. O. Shepard, pp. xvii–xxxi. *Papers of the Southwestern Expedition, Phillips Academy* No. 7. Yale University Press, New Haven, CT.
1936b Speculations on New World Prehistory. In *Essays in Anthropology*, edited by R. Lowie, pp. 143–152. University of California Press, Berkeley.

Kidder, M. A., and Kidder, A. V.
1917 Notes on the Pottery of Pecos. *American Anthropologist* 19:325–360.

King, M.-C., and Wilson, A. C.
1975 Evolution at Two Levels: Molecular Similarities and Biological Differences between Humans and Chimpanzees. *Science* 188:107–116.

Kitts, D. B.
1966 Geologic Time. *Journal of Geology* 74:127–146.
1984 The Names of Species: A Reply to Hull. *Systematic Zoology* 33:112–115.

Klejn, L. S.
1982 Archaeological Typology. *BAR International Series* No. 153. Oxford.

Krieger, A. D.
1944 The Typological Concept. *American Antiquity* 9:271–288.

1947 Certain Projectile Points of the Early American Hunters. *Texas Archeological and Paleontological Society, Bulletin* 18:7–27.

Kristiansen, K.
1985 The Place of Chronological Studies in Archaeology: A View from the Old World. *Oxford Journal of Archaeology* 4:251–266.

Kroeber, A. L.
1909 The Archaeology of California. In *Putnam Anniversary Volume*, edited by F. Boas, pp. 1–42. Stechert, New York.
1916a Zuñi Culture Sequences. *National Academy of Sciences, Proceedings* 2:42–45.
1916b Zuñi Potsherds. *American Museum of Natural History, Anthropological Papers* 18(1): 1–37.
1919 On the Principle of Order in Civilization as Exemplified by Changes of Fashion. *American Anthropologist* 21:235–263.
1925 Archaic Culture Horizons in the Valley of Mexico. *University of California, Publications in American Archaeology and Ethnology* 17(7):373–408.
1927 Coast and Highland in Prehistoric Peru. *American Anthropologist* 29:625–653.
1931a Historical Reconstruction of Culture Growths and Organic Evolution. *American Anthropologist* 33:149–156.
1931b The Culture-Area and Age–Area Concepts of Clark Wissler. In *Methods in Social Science*, edited by S. A. Rice, pp. 248–265. University of Chicago Press, Chicago.
1944 Peruvian Archaeology in 1942. *Viking Fund Publications in Anthropology* No. 4.

Kroeber, A. L., and Strong, W. D.
1924a The Uhle Collections from Chincha. *University of California, Publications in American Archaeology and Ethnology* 21(1):1–54.
1924b The Uhle Pottery Collections from Ica. *University of California, Publications in American Archaeology and Ethnology* 21(2):57–94.

Laufer, B.
1913 The Relation of Archeology to Ethnology: Remarks. *American Anthropologist* 15:573–577.

Leaf, M. L.
1979 *Man, Mind, and Science: A History of Anthropology*. Columbia University Press, New York.

LeBlanc, S. A.
1975 Micro-Seriation: A Method for Fine Chronologic Differentiation. *American Antiquity* 40: 22–38.

Lehmann-Hartleben, K.
1943 Thomas Jefferson, Archaeologist. *American Journal of Archaeology* 47:161–163.

Leonard, R. D., and Jones, G. T.
1987 Elements of an Inclusive Evolutionary Model for Archaeology. *Journal of Anthropological Archaeology* 6:199–219.
1989 (Editors). *Quantifying Diversity*. Cambridge University Press, Cambridge, England.

LeTourneau, P. D.
1998 The "Folsom Problem." In *Unit Issues in Archaeology*, edited by A. F. Ramenofsky and A. Steffen, pp. 52–73. University of Utah Press, Salt Lake City.

Lewin, R.
1982 *Thread of Life: The Smithsonian Looks at Evolution*. Smithsonian Books, Washington, DC.

Lewontin, R. C.
1974 *The Genetic Basis of Evolutionary Change*. Columbia University Press, New York.

Libby, W. F.
1955 *Radiocarbon Dating*, 2nd edition. University of Chicago Press, Chicago.

Lipe, W. D.
1964 Comment on Dempsey and Baumhoff's "The Statistical Use of Artifact Distributions to Establish Chronological Sequence." *American Antiquity* 30:103–104.

Lipo, C., Madsen, M., Dunnell, R. C., and Hunt, T.
1997 Population Structure, Cultural Transmission, and Frequency Seriation. *Journal of Anthropological Archaeology* 16:301–333.
Lowe, J. J. (editor)
1997 *Radiocarbon Dating: Recent Applications and Future Potential.* Wiley, New York.
Lyman, R. L., and Fox, G. L.
1989 A Critical Evaluation of Bone Weathering as an Indication of Bone Assemblage Formation. *Journal of Archaeological Science* 16:293–317.
Lyman, R. L., and O'Brien, M. J.
1997 The Concept of Evolution in Early Twentieth-Century Americanist Archaeology. In *Rediscovering Darwin: Evolutionary Theory and Archeological Explanation*, edited by C. M. Barton and G. A. Clark, pp. 21–48. *American Anthropological Association, Archeological Papers* No. 7.
1998 The Goals of Evolutionary Archaeology: History and Explanation. *Current Anthropology* 39:615–652.
1999 Americanist Stratigraphic Excavation and the Measurement of Culture Change. *Journal of Archaeological Method and Theory* 6:55–108.
Lyman, R. L., O'Brien, M. J., and Dunnell, R. C.
1997a (Editors). *Americanist Culture History: Fundamentals of Time, Space, and Form.* Plenum Press, New York.
1997b *The Rise and Fall of Culture History.* Plenum Press, New York.
Lyman, R. L., Wolverton, S., and O'Brien, M. J.
1998 Seriation, Superposition, and Interdigitation: A History of Americanist Graphic Depictions of Culture Change. *American Antiquity* 63:239–261.
McGregor, J. C.
1941 *Southwestern Archaeology.* Wiley, New York.
1965 *Southwestern Archaeology*, 2nd edition. University of Illinois Press, Urbana.
McKern, W. C.
1934 *Certain Culture Classification Problems in Middle Western Archaeology.* National Research Council, Committee on State Archaeological Surveys, Washington, DC.
1937 Certain Culture Classification Problems in Middle Western Archaeology. In The Indianapolis Archaeological Conference, pp. 70–82. *National Research Council, Committee on State Archaeological Surveys, Circular* 17.
1939 The Midwestern Taxonomic Method as an Aid to Archaeological Culture Study. *American Antiquity* 4:301–313.
Mallory, V. S.
1970 Biostratigraphy—A Major Basis for Paleontologic Correlation. In *Proceedings of the North American Paleontological Convention*, edited by E. L. Yochelson, pp. 553–566. Allen Press, Lawrence, KS.
Marquardt, W. H.
1978 Advances in Archaeological Seriation. In *Advances in Archaeological Method and Theory*, Vol. 1, edited by M. B. Schiffer, pp. 257–314. Academic Press, New York.
Mayr, E.
1959 Darwin and the Evolutionary Theory in Biology. In *Evolution and Anthropology: A Centennial Appraisal*, edited by B. J. Meggers, pp. 1–10. Washington Anthropological Association, Washington, DC.
1963 *Animal Species and Evolution.* Harvard University Press, Cambridge, MA.
1980 Prologue: Some Thoughts on the History of the Evolutionary Synthesis. In *The Evolutionary Synthesis: Perspectives on the Unification of Biology*, edited by E. Mayr and W. B. Provine, pp. 1–48. Harvard University Press, Cambridge, MA.
1981 Biological Classification: Toward a Synthesis of Opposing Methodologies. *Science* 214: 510–516.

1987 The Ontological Status of Species: Scientific Progress and Philosophical Terminology. *Biology and Philosophy* 2:145–166.

Mehringer, Jr., P. J., and Foit, Jr., F. F.

1990 Volcanic Ash Dating of the Clovis Cache at East Wenatchee, Washington. *National Geographic Research* 6:495–503.

Meighan, C. W.

1959 A New Method for the Seriation of Archaeological Collections. *American Antiquity* 25: 203–211.

1966 *Archaeology: An Introduction.* Chandler, San Francisco.

1977 Recognition of Short Time Periods through Seriation. *American Antiquity* 42:628–629.

Meltzer, D. J.

1979 Paradigms and the Nature of Change in American Archaeology. *American Antiquity* 44: 644–657.

1983 The Antiquity of Man and the Development of American Archaeology. In *Advances in Archaeological Method and Theory*, Vol. 6, edited by M. B. Schiffer, pp. 1–51. Academic Press, New York.

1985 North American Archaeology and Archaeologists 1879–1934. *American Antiquity* 50: 249–260.

1991 On "Paradigms" and "Paradigm Bias" in Controversies over Human Antiquity in America. In *The First Americans: Search and Research*, edited by T. Dillehay and D. J. Meltzer, pp. 13–49. CRC Press, Boca Raton, FL.

Meltzer, D. J., and Dunnell, R. C. (Editors)

1992 *The Archaeology of William Henry Holmes.* Smithsonian Institution Press, Washington, DC.

Mercer, H. C.

1897 Researches upon the Antiquity of Man in the Delaware Valley and the Eastern United States. *University of Pennsylvania Series in Philology, Literature, and Archaeology* 6.

Michael, H. N., and Ralph, E. K. (Editors)

1971 *Dating Techniques for the Archaeologist.* MIT Press, Cambridge, MA.

Michaels, G.

1996 Dating the Past. In *The Oxford Companion to Archaeology*, edited by B. M. Fagan, pp. 168–169. Oxford University Press, New York.

Michels, J. W.

1973 *Dating Methods in Archaeology.* Seminar Press, New York.

Moore, R. C., Lalicker, C. G., and Fisher, A. G.

1952 *Invertebrate Fossils.* McGraw-Hill, New York.

Morgan, L. H.

1877 *Ancient Society.* Holt, New York.

Neff, H.

1992 Ceramics and Evolution. In *Archaeological Method and Theory*, Vol. 4, edited by M. B. Schiffer, pp. 141–193. University of Arizona Press, Tucson.

Neiman, F.

1995 Stylistic Variation in Evolutionary Perspective: Inferences from Decorative Diversity and Interassemblage Distance in Illinois Woodland Ceramic Assemblages. *American Antiquity* 60:7–36.

Nelson, N. C.

1906 Excavation of The Emeryville Shellmound, Being a Partial Report of Exploration for the Dep't. of Anthrop during the Year 1906. In Excavation of the Emeryville Shellmound, 1906: Nels C. Nelson's Final Report, edited by J. M. Broughton, pp. 1–47. *University of California, Archaeological Research Facility, Contributions* No. 54.

1910 The Ellis Landing Shellmound. *University of California, Publications in American Archaeology and Ethnology* 7(5):357–426.

1913 Ruins of Prehistoric New Mexico. *American Museum Journal* 13:62–81.

1914 Pueblo Ruins of the Galisteo Basin, New Mexico. *American Museum of Natural History, Anthropological Papers* 15(1):1–124.

1915 Ancient Cities of New Mexico. *American Museum Journal* 15:389–394.

1916 Chronology of the Tano Ruins, New Mexico. *American Anthropologist* 18:159–180.

1918 Chronology in Florida. *American Museum of Natural History, Anthropological Papers* 22(2):74–103.

1919a Human Culture. *Natural History* 19:131–140.

1919b The Archaeology of the Southwest: A Preliminary Report. *National Academy of Sciences, Proceedings* 5:114–120.

1919c The Southwest Problem. *El Palacio* 6:132–135.

1920 Notes on Pueblo Bonito. *American Museum of Natural History, Anthropological Papers* 27:381–390.

1932 The Origin and Development of Material Culture. *Sigma Xi Quarterly* 20:102–123.

Nesbitt, P. H.

1938 Starkweather Ruin: A Mogollon–Pueblo site in the Upper Gila Area of New Mexico, and Affiliative Aspects of the Mogollon Culture. *Logan Museum Bulletin* No. 6.

Nuttall, Z.

1926 The Aztecs and Their Predecessors in the Valley of Mexico. *American Philosophical Society, Proceedings* 65:245–255.

O'Brien, M. J.

1995 Archaeological Research in the Central Mississippi Valley: Culture History Gone Awry. *The Review of Archaeology* 16:23–36.

1996a *Paradigms of the Past: The Story of Missouri Archaeology*. University of Missouri Press, Columbia.

1996b (Editor) *Evolutionary Archaeology: Theory and Application*. University of Utah Press, Salt Lake City.

1998 Sloan: A Dalton-Age Occupation of Northeastern Arkansas. *The Review of Archaeology* 19:16–30.

O'Brien, M. J., and Dunnell, R. C.

1998 A Brief Introduction to the Archaeology of the Central Mississippi River Valley. In *Changing Perspectives on the Archaeology of the Central Mississippi Valley*, edited by M. J. O'Brien and R. C. Dunnell, pp. 1–30. University of Alabama Press, Tuscaloosa.

O'Brien, M. J., and Fox, G. L.

1994a Sorting Artifacts in Space and Time. In *Cat Monsters and Head Pots: The Archaeology of Missouri's Pemiscot Bayou*, by M. J. O'Brien, pp. 25–60. University of Missouri Press, Columbia.

1994b Assemblage Similarities and Differences. In *Cat Monsters and Head Pots: The Archaeology of Missouri's Pemiscot Bayou*, by M. J. O'Brien, pp. 61–93. University of Missouri Press, Columbia.

O'Brien, M. J., and Holland, T. D.

1990 Variation, Selection, and the Archaeological Record. In *Archaeological Method and Theory*, Vol. 2, edited by M. B. Schiffer, pp. 31–79. University of Arizona Press, New York.

1992 The Role of Adaptation in Archaeological Explanation. *American Antiquity* 57:36–59.

1995 Behavioral Archaeology and the Extended Phenotype. In *Expanding Archaeology*, edited by J. M. Skibo, W. H. Walker, and A. E. Nielsen, pp. 143–161. University of Utah Press, Salt Lake City.

O'Brien, M. J., Holland, T. D., Hoard, R. J., and Fox, G. L.

1994 Evolutionary Implications of Design and Performance Characteristics of Prehistoric Pottery. *Journal of Archaeological Method and Theory* 1:259–304.

O'Brien, M. J., and Lyman, R. L.

1998 *James A. Ford and the Growth of Americanist Archaeology*. University of Missouri Press, Columbia.

1999 *Measuring the Flow of Time: The Works of James A. Ford, 1935–1941.* University of Alabama Press, Tuscaloosa, AL.
n.d. *Applying Evolutionary Archaeology: A Systematic Approach.* Plenum Press, New York.
O'Brien, M. J., Lyman, R. L., and Leonard, R. D.
1998 Basic Incompatibilities between Evolutionary and Behavioral Archaeology. *American Antiquity* 63:485–498.
O'Brien, M. J., and Wood, W. R.
1998 *The Prehistory of Missouri.* University of Missouri Press, Columbia.
O'Connell, J. F., and Allen, J.
1998 When Did Humans First Arrive in Greater Australia and Why Is It Important to Know? *Evolutionary Anthropology* 6:132–146.
Osborne, D.
1956 Evidence of the Early Lithic in the Pacific Northwest. *State College of Washington, Research Studies* 24:38–44.
Patterson, T. C.
1963 Contemporaneity and Cross-Dating in Archaeological Interpretation. *American Antiquity* 28:389–392.
Peabody, C.
1904 Explorations of Mounds, Coahoma Co., Mississippi. *Peabody Museum, Papers* 3(2):23–66.
1908 The Exploration of Bushey Cavern, near Cavetown, Maryland. *Phillips Academy, Department of Archaeology, Bulletin* 4(1).
1910 The Exploration of Mounds in North Carolina. *American Anthropologist* 12:425–433.
1913 Excavation of a Prehistoric Site at Tarrin, Department of the Hautes Alpes, France. *American Anthropologist* 15:257–272.
Peabody, C., and Moorehead, W. K.
1904 The Exploration of Jacobs Cavern. *Phillips Academy, Department of Archaeology, Bulletin* 1.
Perino, G.
1968 Guide to the Identification of Certain American Indian Projectile Points. *Oklahoma Anthropological Society, Special Bulletin* No. 3.
1971 Guide to the Identification of Certain American Indian Projectile Points. *Oklahoma Anthropological Society, Special Bulletin* No. 4.
1985 *Selected Preforms, Points and Knives of the North American Indians*, Vol. 1. Points and Barbs Press, Idabel, OK.
1991 *Selected Preforms, Points and Knives of the North American Indians*, Vol. 2. Points and Barbs Press, Idabel, OK.
Petrie, W. M. F.
1899a Sequences in Prehistoric Remains. *Royal Anthropological Institute of Great Britain and Ireland, Journal* 29:295–301.
1899b Prof. Petrie's Address. *Egypt Exploration Fund, Report, 1898* 9:25–29.
1901 Diospolis Parva. *Egypt Exploration Fund, Memoir* No. 20.
Phillips, P.
1958 Application of the Wheat–Gifford–Wasley Taxonomy to Eastern Ceramics. *American Antiquity* 24:117–125.
1970 Archaeological Survey in the Lower Yazoo Basin, Mississippi, 1949–1955. *Peabody Museum of Archaeology and Ethnology, Papers* 60.
Phillips, P., Ford, J. A., and Griffin, J. B.
1951 Archaeological Survey in the Lower Mississippi Valley, 1940–1947. *Peabody Museum of Archaeology and Ethnology, Papers* 25.
Phillips, P., and Willey, G. R.
1953 Method and Theory in American Archaeology: An Operational Basis for Culture-Historical Integration. *American Anthropologist* 55:615–633.

Pitt-Rivers, A. L.-F.
1870 Primitive Warfare, Sec. III; On the Resemblance of the Weapons of Early Races, Their Variations, Continuity and Development of Form—Metal Period. *Royal United Service Institute, Journal* 13:509–539.
1875 On the Principles of Classification Adopted in the Arrangement of His Anthropological Collection, Now Exhibited in the Bethnal Green Museum. *Anthropological Institute of Great Britain and Ireland, Journal* 4:293–308.
Plog, F. T.
1974 *The Study of Prehistoric Change*. Academic Press, New York.
Plog, S., and Hantman, J. L.
1986 Multiple Regression Analysis as a Dating Method in the American Southwest. In *Spatial Organization and Exchange*, edited by S. Plog, pp. 87–113. Southern Illinois University Press, Carbondale.
1990 Chronology Construction and the Study of Prehistoric Change. *Journal of Field Archaeology* 17:439–456.
Praetzellis, A.
1993 The Limits of Arbitrary Excavation. In *Practices of Archaeological Stratigraphy*, edited by E. C. Harris, M. R. Brown III, and G. J. Brown, pp. 68–86. Academic Press, London.
Ramenofsky, A. F.
1998 The Illusion of Time. In *Unit Issues in Archaeology: Measuring Time, Space, and Material*, edited by A. F. Ramenofsky and A. Steffen, pp. 74–84. University of Utah Press, Salt Lake City.
Rands, R. L.
1961 Elaboration and Invention in Ceramic Traditions. *American Antiquity* 26:331–340.
Rastall, R. H.
1956 Geology. *Encyclopaedia Britannica* 10:168. Encyclopaedia Britannica, Chicago.
Rathje, W. L., and Schiffer, M. B.
1982 *Archaeology*. Harcourt Brace Jovanovich, New York.
Rau, C.
1876 The Archaeological Collections of the United States National Museum in Charge of the Smithsonian. *Smithsonian Contributions to Knowledge* 22(4).
Read, D. W.
1979 The Effective Use of Radiocarbon Dates in the Seriation of Archaeological Sites. In *Radiocarbon Dating*, edited by R. Berger and H. E. Suess, pp. 89–94. University of California Press, Berkeley.
Reid, K. C.
1984 Fire and Ice: New Evidence for the Production and Preservation of Late Archaic Fiber-tempered Pottery in the Mid-Latitude Lowlands. *American Antiquity* 49:55–76.
Reiter, P.
1938 Review of *Handbook of Northern Arizona Pottery Wares* by H. S. Colton and L. L. Hargrave. *American Anthropologist* 40:489–491.
Renfrew, C.
1973 *Before Civilization: The Radiocarbon Revolution and Prehistoric Europe*. Knopf, New York.
Roberts, F. H. H., Jr.
1935 A Folsom Complex: Preliminary Report on Investigations at the Lindenmeier Site in Northern Colorado. *Smithsonian Miscellaneous Collections* 94(4).
Rohn, A. H.
1973 The Southwest and Intermontane West. In *The Development of North American Archaeology*, edited by J. E. Fitting, pp. 185–211. Anchor Press, Garden City, NY.
Roth, E., and Poty, B. (Editors)
1989 *Nuclear Methods of Dating*. Kluwer, Dordrecht, Netherlands.

Rouse, I. B.
1939 Prehistory in Haiti, A Study in Method. *Yale University Publications in Anthropology* No. 21.
1953 The Strategy of Culture History. In *Anthropology Today*, edited by A. L. Kroeber, pp. 57–76. University of Chicago Press, Chicago.
1955 On the Correlation of Phases of Culture. *American Anthropologist* 57:713–722.
1967 Seriation in Archaeology. In *American Historical Anthropology: Essays in Honor of Leslie Spier*, edited by C. L. Riley and W. W. Taylor, pp. 153–195. Southern Illinois University Press, Carbondale.
Rowe, J. H.
1959 Archaeological Dating and Cultural Process. *Southwestern Journal of Anthropology* 15: 317–324.
1961 Stratigraphy and Seriation. *American Antiquity* 26:324–330.
1962a Alfred Louis Kroeber. *American Antiquity* 27:395–415.
1962b Stages and Periods in Archaeological Interpretation. *Southwestern Journal of Anthropology* 18:40–54.
1962c Worsaae's Law and the Use of Grave Lots for Archaeological Dating. *American Antiquity* 28:129–137.
Rudwick, M.
1996 Cuvier and Brongniart, William Smith, and the Reconstruction of Geohistory. *Earth Sciences History* 15:25–36.
Sapir, E.
1916 Time Perspective in Aboriginal American Culture, A Study in Method. *Canada Department of Mines, Geological Survey, Memoir* 90.
Sassaman, K.
1993 *Early Pottery in the Southeast: Tradition and Innovation in Cooking Technology*. University of Alabama Press, Tuscaloosa.
Schenck, W. E.
1926 The Emeryville Shellmound: Final Report. *University of California, Publications in American Archaeology and Ethnology* 23(3):147–282.
Schiffer, M. B.
1972 Archaeological Context and Systemic Context. *American Antiquity* 37:156–165.
1987 *Formation Processes of the Archaeological Record*. University of New Mexico Press, Albuquerque.
Schwartz, S. P.
1981 Natural Kinds. *Behavioral and Brain Sciences* 4:301–302.
Schweingruber, F. H.
1989 *Tree Rings: Basics and Applications of Dendrochronology*. Kluwer, Dordrecht, Netherlands.
Setzler, F. M.
1933 Hopewell Type Pottery from Louisiana. *Washington Academy of Sciences, Journal* 23:149–153.
Simons, D. D., Layton, T. N., and Knudson, R.
1985 A Fluted Point from the Mendocino County Coast, California. *Journal of California and Great Basin Anthropology* 7:260–269.
Simpson, G. G.
1943 Criteria for Genera, Species, and Subspecies in Zoology and Paleozoology. *New York Academy of Sciences, Annals* 44:145–178.
Skibo, J. M., Schiffer, M. B., and Reid, K. C.
1989 Organic-Tempered Pottery: An Experimental Study. *American Antiquity* 54: 122–146.
Smith, A. B.
1994 *Systematics and the Fossil Record: Documenting Evolutionary Patterns*. Blackwell, London.
Sober, E.
1980 Evolution, Population Thinking, and Essentialism. *Philosophy of Science* 47:350–383.

1984 *The Nature of Selection: Evolutionary Theory in Philosophical Focus*. MIT Press, Cambridge, MA.

Spaulding, A. C.

1949 Cultural and Chronological Classification in the Plains Area. *Plains Archaeological Conference Newsletter* 2(2):3–5.

1953 Statistical Techniques for the Discovery of Artifact Types. *American Antiquity* 18:305–313.

1954 Reply to Ford. *American Antiquity* 19:391–393.

1960 The Dimensions of Archaeology. In *Essays in the Science of Culture in Honor of Leslie A. White*, edited by G. E. Dole and R. L. Carneiro, pp. 437–456. Crowell, New York.

Spier, L.

1913 Results of an Archeological Survey of the State of New Jersey. *American Anthropologist* 15:675–679.

1916a New Data on the Trenton Argillite Culture. *American Anthropologist* 18:181–189.

1916b Review of *A Pre-Lenape Site in New Jersey* by E. W. Hawkes and R. Linton. *American Anthropologist* 18:564–566.

1917a An Outline for a Chronology of Zuñi Ruins. *American Museum of Natural History, Anthropological Papers* 18(3):207–331.

1917b Zuñi Chronology. *National Academy of Sciences, Proceedings* 3:280–283.

1918a Notes on Some Little Colorado Ruins. *American Museum of Natural History, Anthropological Papers* 18(4):333–362.

1918b The Trenton Argillite Culture. *American Museum of Natural History, Anthropological Papers* 22(4):167–226.

1919 Ruins in the White Mountains, Arizona. *American Museum of Natural History, Anthropological Papers* 18(5):363–388.

1931 N. C. Nelson's Stratigraphic Technique in the Reconstruction of Prehistoric Sequences in Southwestern America. In *Methods in Social Science*, edited by S. A. Rice, pp. 275–283. University of Chicago Press, Chicago.

Stanley, S. M.

1979 *Macroevolution: Pattern and Process*. Freeman, San Francisco.

1981 *The New Evolutionary Timetable*. Basic Books, New York.

Stein, J. K.

1990 Archaeological Stratigraphy. In Archaeological Geology of North America, edited by N. P. Lasca and J. Donahue. *Geological Society of America, Centennial Special Volume* 4: 513–523.

Sterns, F. H.

1915 A Stratification of Cultures in Nebraska. *American Anthropologist* 17:121–127.

Stevens, S. S.

1946 On the Theory of Scales of Measurement. *Science* 103:677–680.

Steward, J. H.

1929 Diffusion and Independent Invention: A Critique of Logic. *American Anthropologist* 31: 491–495.

1941 Review of *Prehistoric Culture Units and Their Relationships in Northern Arizona* by H. S. Colton. *American Antiquity* 6:366–367.

1942 The Direct Historical Approach to Archaeology. *American Antiquity* 7:337–343.

1954 Types of Types. *American Anthropologist* 56:54–57.

Strong, W. D.

1925 The Uhle Pottery Collections from Ancon. *University of California, Publications in American Archaeology and Ethnology* 21(4):135–190.

1935 An Introduction to Nebraska Archaeology. *Smithsonian Miscellaneous Collections* 93(10).

1940 From History to Prehistory in the Northern Great Plains. In Essays in Historical Anthropology of North America. *Smithsonian Miscellaneous Collections* 100:353–394.

1952 The Value of Archeology in the Training of Professional Anthropologists. *American Anthropologist* 54:318–321.

Strong, W. D., and Evans, Jr., C.

1952 *Cultural Stratigraphy in the Virú Valley, Northern Peru: The Formative and Florescent Epochs*. Columbia University Press, New York.

Stuiver, M., and Becker, B.

1986 A Calibration Curve for the Radiocarbon Timescale. *Radiocarbon* 28:863–910.

Stuiver, M., and Kra, R. (Editors)

1986 Calibration Issue: 12th International Radiocarbon Conference. *Radiocarbon* 28.

Stuiver, M., and Reimer, P. J.

1986 A Computer Program for Radiocarbon Age Calibration. *Radiocarbon* 28:1022–1030.

1993 Extended ^{14}C Data Base and Revised CALIB 3.0 ^{14}C Age Calibration Program. *Radiocarbon* 35:215–230.

Suhm, D. A., and Jelks, E. B.

1962 Handbook of Texas Archeology: Type Descriptions. *Texas Archeological Society*, Special Publication No. 1.

Suhm, D. A., Krieger, A. D., and Jelks, E. B.

1954 An Introductory Handbook of Texas Archeology. *Texas Archeological Society*, Bulletin 25.

Taylor, R. E.

1985 The Beginnings of Radiocarbon Dating in *American Antiquity*: A Historical Perspective. *American Antiquity* 50:309–325.

1987 *Radiocarbon Dating: An Archaeological Perspective*. Academic Press, Orlando.

1996 Radiocarbon Dating: The Continuing Revolution. *Evolutionary Anthropology* 4:169–181.

Taylor, R. E., and Aitken, M. J. (Editors)

1997 *Chronometric Dating in Archaeology*. Plenum Press, New York.

Taylor, R. E., Haynes, Jr., C. V., and Stuiver, M.

1996 Clovis and Folsom Age Estimates: Stratigraphic Context and Radiocarbon Calibration. *Antiquity* 70:515–525.

Taylor, R. E., Long, A., and Kra, R. S. (Editors)

1992 *Radiocarbon Dating after Four Decades: An Interdisciplinary Perspective*. Springer-Verlag, Berlin.

Taylor, W. W.

1948 A Study of Archeology. *American Anthropological Association, Memoir* 69.

1954 Southwestern Archaeology, Its History and Theory. *American Anthropologist* 56:561–575.

Thomas, C.

1884 Who Were the Mound Builders? *The American Antiquarian* 6:90–99.

1891 Catalogue of Prehistoric Works East of the Rocky Mountains. *Bureau of Ethnology, Bulletin* 12.

1894 Report on the Mound Explorations of the Bureau of Ethnology. *Bureau of Ethnology, Annual Report* 12:3–742.

Thomas, D. H.

1971 *Prehistoric Subsistence-Settlement Patterns of the Reese River Valley, Central Nevada*. PhD dissertation, University of California, Davis.

1983 The Archaeology of Monitor Valley 2. Gatecliff Shelter. *American Museum of Natural History, Anthropological Papers* 59(1).

1998 *Archaeology*, 3rd edition. Harcourt Brace, Fort Worth, Texas.

Thompson, R. H. (Editor)

1956 An Archaeological Approach to the Study of Cultural Stability. In Seminars in Archaeology: 1955. *Society for American Archaeology, Memoirs* 11:31–57.

Thomsen, C. J.

1848 (orig. 1836) *A Guide to Northern Antiquities*. London.

Toulmin, S., and Goodfield, J.
1965 *The Discovery of Time*. University of Chicago Press, Chicago.
Tozzer, A. M.
1926 Chronological Aspects of American Archaeology. *Massachusetts Historical Society, Proceedings* 59:283–292.
1937 Mexico, Central and South America [review of G. C. Vaillant 1930, 1931, 1935a, 1935b and S. B. Vaillant and G. C. Vaillant 1934]. *American Anthropologist* 39:338–340.
Trigger, B.
1978 *Time and Traditions*. Columbia University Press, New York.
1989 *A History of Archaeological Thought*. Cambridge University Press, Cambridge, England.
Tylor, E. B.
1871 *Primitive Culture*. Murray, London.
Uhle, F. M.
1902 Types of Culture in Peru. *American Anthropologist* 4:753–759.
1903 *Pachacamac*. University of Pennsylvania Press, Philadelphia.
1907 The Emeryville Shellmound. *University of California, Publications in American Archaeology and Ethnology* 7:1–107.
Vaillant, G. C.
1930 Excavations at Zacatenco. *American Museum of Natural History, Anthropological Papers* 32(1):1–198.
1931 Excavations at Ticomán. *American Museum of Natural History, Anthropological Papers* 32(2):199–432.
1935a Early Cultures of the Valley of Mexico: Results of the Stratigraphical Project of the American Museum of Natural History in the Valley of Mexico, 1928–1933. *American Museum of Natural History, Anthropological Papers* 35(3):281–328.
1935b Excavations at El Arbolillo. *American Museum of Natural History, Anthropological Papers* 35(2):137–279.
1936 The History of the Valley of Mexico. *Natural History* 38:324–328.
1937 History and Stratigraphy in the Valley of Mexico. *Scientific Monthly* 44:307–324.
Vaillant, S., and Vaillant, G. C.
1934 Excavations at Gualupita. *American Museum of Natural History, Anthropological Papers* 35(1):1–135.
Van Riper, A. B.
1993 *Men among the Mammoths: Victorian Science and the Discovery of Human Prehistory*. University of Chicago Press, Chicago.
Volk, E.
1911 The Archaeology of the Delaware Valley. *Peabody Museum of American Archaeology and Ethnology, Papers* 5.
Wagner, G., and van den Haute, P.
1992 *Fission-Track Dating*. Kluwer, Dordrecht, Netherlands.
Walker, S. T.
1883 The Aborigines of Florida. *Smithsonian Institution, Annual Report* (1881), pp. 677–680.
Watson, P. J.
1990 Trend and Tradition in Southeastern Archaeology. *Southeastern Archaeology* 9:43–54.
Webb, W. S., and DeJarnette, D. L.
1942 An Archaeological Survey of Pickwick Basin in the Adjacent Portions of the States of Alabama, Mississippi, and Tennessee. *Bureau of American Ethnology, Bulletin* 129.
Wedel, W. R.
1938 The Direct-Historical Approach in Pawnee Archaeology. *Smithsonian Miscellaneous Collections* 97(7).

1940 Culture Sequence in the Central Great Plains. In Essays in Historical Anthropology of North America, pp. 291–352. *Smithsonian Miscellaneous Collections* 100.

Whallon, R., and Brown, J. A. (Editors)

1982 *Essays on Archaeological Typology*. Center for American Archaeology Press, Evanston, IL.

Wheat, J. B., Gifford, J. C., and Wasley, W. W.

1958 Ceramic Variety, Type Cluster, and Ceramic System in Southwestern Pottery Analysis. *American Antiquity* 24:34–47.

Wheeler, M.

1956 *Archaeology from the Earth*. Penguin Books, Baltimore.

Willey, G. R.

1936 *A Survey of Methods and Problems in Archaeological Excavation, with Special Reference to the Southwest*. M.A. thesis, University of Arizona.

1938 Time Studies: Pottery and Trees in Georgia. *Society for Georgia Archaeology, Proceedings* 1:15–22.

1939 Ceramic Stratigraphy in a Georgia Village Site. *American Antiquity* 5:140–147.

1945 Horizon Styles and Pottery Traditions in Peruvian Archaeology. *American Antiquity* 10:49–56.

1951 The Chavín Problem: A Review and Critique. *Southwestern Journal of Anthropology* 7: 103–144.

1953 Archaeological Theories and Interpretation: New World. In *Anthropology Today*, edited by A. L. Kroeber, pp. 361–385. University of Chicago Press, Chicago.

1968 One Hundred Years of American Archaeology. In *One Hundred Years of Anthropology*, edited by J. O. Brew, pp. 26–53. Harvard University Press, Cambridge, MA.

1988 *Portraits in American Archaeology: Remembrances of Some Distinguished Americanists*. University of New Mexico Press, Albuquerque.

Willey, G. R., and Phillips, P.

1958 *Method and Theory in American Archaeology*. University of Chicago Press, Chicago.

Willey, G. R., and Sabloff, J. A.

1993 *A History of American Archaeology*, 3rd edition. Freeman, New York.

Willey, G. R., and Woodbury, R. B.

1942 A Chronological Outline for the Northwest Florida Coast. *American Antiquity* 7:232–254.

Willis, J. C.

1922 *Age and Area, a Study in Geographical Distribution and the Origin of Species*. Cambridge University Press, Cambridge, England.

Wissler, C.

1915 Explorations in the Southwest by the American Museum. *American Museum Journal* 15: 395–398.

1916a Correlations between Archeological and Culture Areas in the American Continents. In *Holmes Anniversary Volume: Anthropological Essays*, edited by F. W. Hodge, pp. 481–490, Wsahington, DC.

1916b The Application of Statistical Methods to the Data on the Trenton Argillite Culture. *American Anthropologist* 18:190–197.

1916c The Genetic Relations of Certain Forms in American Aboriginal Art. *National Academy of Sciences, Proceedings* 2:224–226.

1917a *The American Indian*. McMurtrie, New York.

1917b The New Archaeology. *American Museum Journal* 17:100–101.

1919 General Introduction. *American Museum of Natural History, Anthropological Papers* 18: iii–ix.

1921 Dating Our Prehistoric Ruins. *Natural History* 21:13–26.

Woodbury, R. B.

1960a Nels C. Nelson and Chronological Archaeology. *American Antiquity* 25:400–401.

1960b Nelson's Stratigraphy. *American Antiquity* 26:98–99.
1973 *Alfred V. Kidder*. Columbia University Press, New York.

Worsaae, J. J. A.
1849 *The Primeval Antiquities of Denmark*. Parker, London.

Wyman, J.
1868 An Account of Some Kjoekkenmoeddings, or Shell-Heaps, in Maine and Massachusetts. *The American Naturalist* 1:561–584.
1875 Fresh-Water Shell Mounds of the St. John's River, Florida. *Peabody Academy of Science, Memoir* 4.

Index

Absolute dating methods, v, 6, 8, 9, 12, 15; *see also* Amino acid racemization; Archaeomagnetism; Dendrochronology; Electron spin resonance; Fission track dating; Potassium–argon dating; Radiocarbon; Thermoluminescence
Acculturation, 218
Age, 6, 25
Age–area concept, 82, 222
Aleutian Islands, 153
American Indian, The, 174
Americanist archaeology, vi, vii, 1, 3, 4, 7, 22, 24, 32, 34, 56, 57, 59, 61, 64, 109, 112, 149, 150, 152, 154, 156, 166, 173, 187, 188, 196, 207, 218, 219, 221, 222, 225; *see also* Culture, history; Processual archaeology
American Museum of Natural History, 26, 160, 162, 165, 166, 167, 188, 192, 222
Amino acid racemization, 6
An Introductory Handbook of Texas Archeology, 40
Analogues, vii, 75, 80, 82, 83, 84, 89, 91, 93, 97, 99, 214, 224; *see also* Homologues
Ancient Society, 153
Antiquarianism, 1, 2
Arbitrary levels, 147, 149, 156, 162, 168, 170, 197, 200, 220
Archaeological record
  minimal time depth, vi, 2, 3, 157, 162, 165, 21, 222
Archaeomagnetism, 6
Archetypes, 104, 105, 106
Arrow points, 41
Ascher, Robert, 100
Assemblage, 25, 110, 114, 117, 131
Attributes; *see also* Character states
  as chronological markers, 38, 52
  functional, 38, 40, 89
  in seriation, 61, 131
  in type construction, 40, 41, 53, 54, 57, 58, 96, 106, 127, 209, 212, 215
Atzcapotzalco (Mexico), 157, 174, 192
Babbage, Charles, 12
Battleship-shaped curves, 125, 128, 183
Baumhoff, Martin, 119, 129, 130, 136
Baytown Plain, 51, 218
Beach dating, 6
Bell Plain, 36, 37, 38, 40, 89
Biers, William, 109
Big bang, 20
Bighorn Basin (Wyoming), 70
Binford, Lewis, 134, 135
Biostratigraphy, 185, 186, 188, 190, 191, 193, 197, 199, 209, 214
Bison, 188, 189, 190
Blackwater Draw (New Mexico), 189
Boas, Franz, 26, 64, 157, 222
Bone weathering, 19
Bradley, Bruce, 106
Brainerd, George, 52, 53, 54, 57, 116, 119
Braun, David, 134
Breuil, Henri, 161
Brew, J. O., 58, 99, 100, 224
Brixham Cave (England), 152
Brongniart, Alexandre, 185
Browman, David, 150, 157; *see also* Givens, Douglas
Bureau of (American) Ethnology, 2, 3, 33, 152, 158

Carnegie Institution, 192
Castillo Cave (Spain), 161
Cause, 218
Ceramic complexes, 34
Ceramic continuum, 42, 44
Ceramic stratigraphy: *see* Percentage stratigraphy
Change
  viewed continuously, 183, 188, 198, 203, 215, 225
  viewed discontinuously, 183, 188, 203, 225
Character states, 73, 75, 76, 77, 78, 94, 97, 101, 106, 109; *see also* Symplesiomorphies; Synapomorphies
  differences in rates of change, 80

Chemical change in sediments, 6
Chinese civilization, 18
Chronocline: *see* Chronospecies
Chronological types: *see* Historical, types
Chronology, 1, 4, 5
  conflation of measures, 14
Chronometric dating: *see* Absolute dating methods
Chronospecies, 68, 70, 71, 74, 134
Clades, 77, 118
Cladistical analysis, 77, 78; *see also* Stratocladistics
Cladogram, 77
Classes, 34, 75, 76, 89, 93, 96, 99, 110, 115, 116, 117, 118, 127, 131, 209
Classification systems, 32, 58
Clovis points, 41, 102, 103, 104, 105, 106, 189, 190, 191
Coles Creek complex, 201, 203, 205
Collins, Henry B., 199, 200
Colorado Museum of Natural History, 188
Colton, Harold, 96, 97, 99, 224
Columbia University, 26
Common sense, 217, 218, 223
Complex linking, 209
Complex markers: *see* Index fossils
Component, 25, 172, 179, 180, 183
Conference on Southern Pre-History, 199
Continuous occurrence, 126
Convergence, 82, 83
Contemporaneity, 120, 123
Cowgill, George, 114, 115, 125, 126, 127, 129
Cross dating, v, 6, 7, 13, 87, 137, 190, 209, 213; *see also* Index fossils
Cultural norms, 52, 55; *see also* Pottery types, creation versus discovery of
Cultural stream, 52, 55, 99, 114, 121, 223, 224
Culture, 200, 204
  change, vi, 10, 80, 82, 115, 165, 174, 182, 200, 212, 219, 225, 226
  history, vi, vii, 7, 10, 34, 55, 57, 60, 93, 99, 121, 162, 174, 214
  traits, 4, 82, 83, 84, 187, 213, 219, 220, 222, 224, 226
Culture-area concept, 222
Cultures, 144, 174, 222, 223
Cuvier, Georges, 185

Dall, William Healey, 153, 154, 162
Dalton points, 41, 101, 102, 103, 104, 105, 106, 107
Darwin, Charles, 65, 74, 93
Date, 6
Dating, 5; *see also* Target event
  methods, 13, 14, 59, 225
    direct, 11, 12, 61
    indirect, 11, 12, 61, 145
Davis, Edwin, 12
Deasonville (Mississippi), 201
Deasonville complex, 201, 205
Decoration, 24, 28, 48, 89, 94, 133, 172, 193, 202, 204
Decoration complexes, 200, 203, 204, 205, 209, 212, 213
  mixing of, 209, 210
Decoration types, 40, 203, 204, 211, 212
Deetz, James, 46, 126; *see also* Dethlefsen, Edwin
Delaware River, 165, 166
Dempsey, Paul, 119, 129, 130, 136; *see also* Baumhoff, Martin
Dendrochronology, 6, 12, 15, 112, 120, 187, 221, 225; *see also* Babbage, Charles; Douglass, A. E.
Depositional event, 144, 145, 149
Descent with modification, 65, 74, 93; *see also* Heritable continuity
Dethlefsen, Edwin, 46, 126
Diffusion, 46, 118, 196, 209, 211, 213, 214, 218
Dimensional classification: *see* Paradigmatic classification
Dinosaurs, 78
Dionysius Exiguus, 18, 19
Direct historical approach, 199, 220
Doppler effect: *see* Time lag
Douglass, A. E., 12
Drennan, Robert, 134
Dunnell, Robert, 57, 126

Edwards Mound (Mississippi), 154
Egypt, 63, 84, 94
Egyptian civilization, 18
Ehrich, Robert, 144
Einstein, Albert, 20
El Arbolillo (Mexico), 192, 197
Eldredge, Niles, 185, 186, 214
Electron spin resonance, 6
Ellis Landing (California), 160
Emeryville Shellmound (California), 158, 160, 162, 163
Empirical units, 22, 44, 51, 52, 55, 70, 96, 101, 108, 110, 113, 218, 220, 223
Essentialism, 56, 219, 222

Essentialist–materialist paradox, 225, 226
Ethnography, 1
European archaeology, 1, 3, 7, 59
Evans, John, 91, 111, 212
Event, 6
Evolution, vii, 52, 62, 66, 72, 74, 75, 78, 80, 82, 88, 100, 101, 106, 186, 188, 196, 200, 212, 215, 224, 226
Evolutionary archaeology, vii, 101
Excavation techniques, 139, 140, 173, 193, 200

Figgins, Jesse, 191
Figurines, 193, 195, 196, 197, 198, 199
Fission track dating, 6
Flat-past perspective: *see* Archaeological record, minimal time depth
Flint knapping, 106
Folsom (New Mexico), 4, 188, 189, 191
Folsom points, 41, 103, 106, 189, 190, 191
Ford, James A., 25, 33, 34, 35, 36, 40, 41, 42, 44, 46, 51, 52, 53, 54, 55, 56, 57, 58, 61, 70, 80, 99, 100, 115, 119, 121, 123, 124, 126, 181, 182, 183, 199, 200, 201, 204, 205, 206, 207, 209, 211, 213, 218, 225
Fowke, Gerard, 174
Frequencies
  plotted against time, 115
  of traits, 4
  of types, 110, 113, 116, 119, 121, 130, 220, 222

Galisteo Basin (New Mexico), 26, 160, 161, 174, 220
Gamio, Manuel, 157, 158, 170, 174, 192, 222, 226
Gatecliff Shelter (Nevada), 175, 176, 177, 180, 183
Gayton, Anna, 63
Geological Survey of New Jersey, 165
Gibson, Jon, 207
Gifford, James, 56
Gingerich, Philip, 70
Givens, Douglas, 150, 157
Gladwin, Harold, 96, 224
Goodyear, Albert, 102, 103
Gould, Stephen Jay, 10, 185, 186, 214
Graves, Michael, 120
Great Basin, 175
Gregorian calendar, 18
Griffin, James B., 33, 34, 35, 36, 40, 41, 42, 44, 46, 51, 58, 70, 121, 123, 126, 182, 218
Gulf of Mexico, 48

Handbook of Texas Archeology: Type Descriptions, 40
Hantman, Jeff, 135
Hardaway points, 107
Hardaway Dalton points, 107
Hargrave, Lyndon, 96, 97, 99
Harper, C. W., 185
Harrington, J. C., 134
Harrington, Mark, 154, 156, 162, 163
Harvard University, 2, 168, 191
Haury, Emil, 214, 215
Hay, Clarence C., 192
Heritability, 116, 127
Heritable continuity, 60, 66, 67, 68, 73, 74, 75, 76, 80, 91, 93, 94, 95, 96, 101, 106, 108, 109, 111, 115, 117, 118, 121, 129, 185, 191, 200, 203, 212, 214, 215, 224
  tradition–lineage sense, 117, 118, 127
  type–species sense, 116, 118
Historical
  continuity, 60, 65, 66, 67, 73, 80, 90, 93, 94, 95, 96, 99, 101, 108, 109, 111, 115, 123, 129, 130, 214, 224
  relatedness, 93, 97
  significance test, 40, 54, 180, 221, 224
  types, 24, 25, 26, 28, 30, 57, 58, 111, 116, 117, 123, 131, 187, 196, 226
Hole, Frank, 131; *see also* Shaw, Mary
Holmes, William Henry, 2, 32, 33, 40, 166, 173
Homologues, vii, 75, 82, 83, 84, 89, 91, 93, 99, 101, 106, 108, 109, 214, 224
Horizons, 100, 175, 179, 209, 213, 214
Hubble, Edwin, 20
Humboldt Concave Base points, 180

Ideational units, 21, 22, 51, 52, 55, 56, 60, 70, 96, 108, 110, 113, 114, 117, 118, 198, 204, 205, 207, 209, 218, 220, 223
Immigration, 196
Independent invention, 82, 83, 212
Index fossils, vii, 6, 73, 87, 185, 187, 188, 191, 196, 197, 199, 201, 203, 204, 206, 209, 212, 213, 214, 215, 223, 224, 226
  as indicators of cultures, 213
Inference, 11, 14, 22, 55, 60, 66, 96, 100, 126, 145, 191, 214, 221, 224
Interpreting site history, 140
Interval scale, 9, 18, 132, 134, 136

Jacobs Cavern (Missouri), 154

Jefferson, Thomas, 151, 152
Julian calendar, 19

Kidder, A. V., 33, 62, 63, 95, 97, 101, 109, 110, 112, 144, 158, 162, 167, 168, 170, 171, 187, 188, 192, 200, 212, 220, 221, 223, 225, 225, 226
Krieger, Alex, 26, 180
Kroeber, A. L., 26, 33, 52, 60, 62, 63, 64, 82, 83, 84, 85, 88, 94, 101, 111, 112, 113, 114, 126, 158, 160, 165, 171, 209, 213, 220, 221, 225

Larto Red Filmed, 44
LeBlanc, Stephen, 124, 130, 131
Lemaitre, Georges, 20
Lewontin, Richard, 218
Libby, Willard, 12
Lineages, 38, 52, 65, 67, 68, 70, 71, 72, 73, 75, 78, 99, 101, 109, 114, 129, 131, 224, 226
Linnaean taxonomic system, 68, 100, 118
Lipe, William, 119
Lipo, Carl, 128
Local area, 118, 127, 128
Louisiana–Mississippi pottery sequence, 201

McGregor, John, 173
Mammoth, 189, 190
Marker types: *see* Index fossils
Marksville complex, 200, 201
Marksville Stamped, 46, 48
Marquardt, William, 59
Materialism, 42, 219
Mayan civilization, 18
Mayr, Ernst, 68
Mazique Incised, 44
Meighan, Clement, 115, 130, 183
Meltzer, David J., 3
Mercer, Henry C., 165, 166
Metaphysics: *see* Ontological positions
Mexico, 157, 158, 198, 212, 213
Microseriation, 131
Midwest, 30, 58, 102, 104, 105, 106, 189
Midwestern Taxonomic Method, 223
Mississippi River, 30, 32, 33, 34, 37, 40, 48, 56, 121, 126
Mississippian period, 32, 48
Modern Synthesis, 188
Mohs hardness scale, 9
Molecular clocks, 133
Monitor Valley (Nevada), 175
Monophyletic groups: *see* Clades
Moorehead, Warren K., 2, 154
Morgan, Lewis Henry, 4, 62, 153
Morgan, Thomas H., 100
Moundbuilders, 2
Mulberry Creek Cordmarked, 44
Murchison, Sir Roderick Impey, 185

National Research Council, 199
Natural selection, 67
Natural stratigraphic units, 162, 170, 174, 200, 220
Neeley's Ferry Plain, 36, 37, 38, 40, 89
Neiman, Fraser, 128
Nelson, Nels C., 4, 26, 28, 33, 34, 36, 52, 58, 61, 115, 144, 147, 158, 160, 161, 162, 163, 165, 167, 168, 170, 171, 172, 200, 220, 221, 223, 225, 226
New England, 152
Nominal scale, 8, 17
Northwest, 191
Nuttal, Zelia, 157

Obermaier, Otto, 161
Oceania, 120
Ohio River, 34, 48
Oklahoma Anthropological Society, 41
On the Origin of Species, 93
Ontological positions, 219, 223
Ordinal scale, 9, 17, 18, 25, 134, 145
Osborne, Douglas, 190, 191
Overlapping traits, 115

Paaco (New Mexico), 28
Pajarito Plateau (New Mexico), 112
Paleoindian period, 190
Paradigmatic classification, 96
Parkin Punctated, 44, 46, 48
Paste, 37, 48, 57, 89
Pattern repetition, 186, 199
Peabody, Charles, 154, 156
Peabody Museum (Harvard), 2, 152
Peck Village (Louisiana), 200, 201, 202, 203, 204, 205, 206
Pecos Pueblo (New Mexico), 94, 109, 162, 167, 168, 170, 187, 192
Pelycodus, 70
Percentage stratigraphy, 170, 171, 220
Perino, Gregory, 30, 41
Period, 10

Petrie, W. M. F., 61, 63, 64, 84, 85, 87, 88, 89, 90, 91, 94, 110, 111, 212
Phenograms, 76
Phenotypes, 52, 101, 186
Phillips Academy, 2, 154
Phillips, Philip, 34, 35, 36, 40, 41, 42, 44, 46, 48, 49, 51, 52, 70, 100, 101, 121, 123, 126, 179, 182, 207, 218, 225
Phylogenetic history, 67, 78, 91, 101, 214
Phylogeny, 96, 100, 101
Pipe stems, 134, 135
Pitt-Rivers, A. L.-F., 91, 93, 100
Plains, 3, 104, 106, 191
Plog, Fred, 183
Plog, Stephen, 135; *see also* Hantman, Jeff
Popularity principle, 63, 129, 171, 220
Potassium-argon dating, 6
Pottery types, 32, 42, 53, 193, 200, 214, 221
  as chronological markers, 36, 204
  creation versus discovery of, 35, 44, 51, 52, 53, 54, 55, 56, 57
  default categories, 38
  descriptions versus necessary criteria, 49, 57
  Ford's decoration system of, 207, 209
  geographic distribution, 199, 205, 209
  Gladwin–Colton–Hargrave system of, 96, 97, 98, 99, 100, 101
  index system of, 204, 205, 206
  Kroeber system of, 111, 112, 113, 114
  Phillips, Ford, and Griffin system of, 36, 41, 42, 44
  problems when considering space, 44, 46
  replacement of, 203
  type-variety system of, 48, 56, 198, 207
  use in chronology building, 26, 28, 34, 121
  usefulness of characters in defining, 36, 38
Powell, John Wesley, 2
Presence–absence
  of traits, 4, 76, 163, 226
  of types, 110, 116, 119, 120, 123, 124, 130
Principle of association, 11; *see also* Worsaae, J. J. A.
Processual archaeology, 7, 174, 175
Projectile point guides, 40-41, 104
Projectile point shape, 175
Projectile point types, 30, 32
  as chronological markers, 41
  phylogenetic analysis of, 101, 102, 103, 104, 105, 106, 107
  variation in morphological characteristics, 30, 189
Pueblo San Cristobal (New Mexico), 26, 28, 147, 161, 162, 163, 167
Punctuated equilibrium, 134, 188
Putnam, Frederic Ward, 2, 157

Quaternary Isotope Laboratory, 17

Radiocarbon, v, 6, 9, 12, 13, 15, 17, 102, 120, 137, 190, 225; *see also* Libby, Willard
  association with artifacts, 30, 134
  calibration of dates, 15, 17
Radiocarbon, 15, 17
Radiometric dating, 13, 15, 30, 190, 226
Rates of change, 80, 114, 115, 124, 131, 132, 135
Ratio scale, 19
Rau, Charles, 32
Read, Dwight, 134
Reese River (Nevada), 180
Regression analysis, 134, 135, 136
Reisner, George, 94, 192
Reiter, Paul, 96, 99
Relative dating, v, 12, 13, 25
  methods, v, vi, vii, 6, 8, 9, 12, 13, 24, 58, 108, 120, 137, 183, 226; *see also* Beach dating; Chemical change in sediments; Cross dating; Sediment accumulation; Seriation; Stratigraphy; Terrace dating; Varve dating
Reoccupation of sites, 210
Replication, 73
Reversed stratigraphy: *see* Stratigraphy, mixed
Roberts, Frank H. H., 188
Rodgers Shelter (Missouri), 103, 104
Rouse, Irving, 25, 54, 93, 117, 118, 213, 214
Rowe, John, 61, 112, 119

Sample
  error, 125, 126
  size, 117, 119, 126, 200
Sapir, Edward, 83, 84
Scale of observation, 4
Scales of measurement, 8, 17; *see also* Interval scale; Nominal scale; Ordinal scale; Ratio scale
Schiffer, Michael B., 144
Science, 217, 218, 219, 221
Sedgwick, Adam, 185
Sediment accumulation, 6
Sediments
  age of, 145, 147

Selected Preforms, Points and Knives of the North American Indians, 41
Sequence dating: *see* Seriation, phyletic
Seriation, v, vii, 6, 7, 13, 18, 67, 80, 108, 109, 110, 111, 115, 116, 117, 119, 126, 128, 132, 134, 136, 137, 183, 187, 209, 215, 221, 225, 226
  absolute, 111, 134, 135
  conditions, 125, 126, 127, 128, 129
  conflation of techniques, 59, 60, 61, 62, 63
  definition, 62, 63, 65, 74
  developmental: *see* Seriation, evolutionary
  early use, 62, 64
  evolutionary, 64, 65, 88
  frequency, 7, 59, 60, 63, 64, 85, 109, 110, 111, 112, 114, 115, 116, 118, 119, 121, 123, 124, 126, 128, 129, 130, 131, 132, 133, 136, 137, 220, 221
  occurrence, 59, 60, 64, 109, 110, 111, 114, 116, 118, 119, 120, 121, 125, 126, 130, 132, 133, 137
  phyletic, 59, 60, 63, 64, 65, 66, 67, 84, 85, 87, 91, 93, 94, 95, 101, 108, 109, 110, 111, 112, 114, 133, 196, 212, 220, 221, 223
  requirements, 117, 118, 119
  similiary, 64
Setzler, Frank M., 199, 200
Shared derived characters: *see* Synapomorphies
Shared primitive characters: *see* Symplesiomorphies
Shaw, Mary, 131
Shell mounds, 152, 153, 160, 163
Simpson, George Gaylord, 68, 70
Site formation, 144
Smith, William, 185
Smithsonian Institution, 188, 199
Society for American Archaeology 1955 Seminar, 225
Southeast, 12, 30, 33, 46, 58, 61, 102, 105, 199
Southwest, 4, 7, 12, 34, 40, 58, 62, 96, 104, 106, 115, 120, 158, 161, 167, 187, 191, 221, 222
Spain, 161
Spaulding, Albert, 13, 24, 53, 54, 55, 56, 57, 225
Species, 52, 67, 68, 70, 72, 73, 74, 82, 83; *see also* Chronospecies
Spier, Leslie, 62, 63, 94, 158, 162, 165, 167, 171, 220, 221, 225, 226
Squier, Ephraim, 12; *see also* Davis, Edwin
St. Francis River (Arkansas), 46
Statistics, 54
Steno, Nicolaus, 144
Sterns, Fred H., 157
Steward, Julian, 24, 84, 93, 94, 101
Stone Age Spear and Arrow Points of the Midcontinental and Eastern United States, 41
Stratification, 142, 193
Stratigraphic
  excavation, v, vi, 137, 140, 142, 147, 149, 150, 154, 160, 163, 172, 173, 180, 197, 220, 222, 225, 226
    bread loaf technique, 150, 151, 156, 160, 166
    onion peel technique, 150, 151, 156
  observation: *see* Stratigraphy
  principles, 152
  revolution, vi, 3, 149, 151, 171, 221, 225
Stratigraphy, vii, 6, 7, 13, 14, 28, 34, 61, 62, 73, 78, 82, 94, 112, 130, 133, 140, 142, 144, 152, 154, 157, 168, 172, 175, 183, 200, 213, 221, 226; *see also* Percentage stratigraphy
  mixed, 146, 173, 187, 192, 222, 223
Stratocladistics, 78
Stratum, 142, 144, 146, 147, 149, 154, 158, 170, 173, 174, 175, 176, 177, 183, 185, 187, 188, 192, 193, 198, 215, 222
  as data recovery units, 151, 156
Strong, William Duncan, 63, 221; *see also* Gayton, Anna
Styles, 4, 58, 172, 196, 214
Stylistic change, 109
Sumerian civilization, 18
Superposition, 6, 7, 14, 59, 61, 62, 66, 74, 78, 82, 94, 95, 115, 137, 142, 144, 145, 151, 152, 158, 162, 163, 166, 168, 172, 173, 174, 180, 183, 187, 196, 219, 221, 222, 224, 225
  conflation of strata and age of sediments, 145, 146
Surface treatment, 26, 46, 48, 57
Symplesiomorphies, 76, 78, 89, 91, 97, 106
Synapomorphies, 76, 77, 78, 89, 91, 97, 106, 108, 109

Talus, 177
Tano Ruins (New Mexico), 26, 162
Target event, 11, 25, 26, 145
Taxonomic schemes, vii
Taylor, R. E., 190
Taylor, Walter, 53

Temper, 33
Temporal types, 24
Terrace dating, 6
Thebes points, 30, 51
Theoretical units: *see* Ideational units
Theory, 21, 52, 67, 93, 98, 214, 218, 224
Thermoluminescence, v, 6, 12
Thomas, Cyrus, 2, 12, 152
Thomas, David Hurst, 23, 175, 179, 183
Thomsen, Christian J., 152
Ticoman (Mexico), 192, 193, 196
Time, 217
  continuous view, 9, 10, 60, 61, 62, 73, 137, 188, 220, 223
  as a dimension, 20
  discontinuous view, 9, 10, 60, 61, 62, 73, 180, 188, 198, 199, 212
  measurement of, 14, 15, 223
  nature of, 8, 13, 20, 21, 24, 40, 44, 72, 120, 219
  views of, 10, 11, 14
Time lag, 46
Time's Arrow, Time's Cycle, 10
Time–space continuum, 55, 121, 187
Tozzer, Alfred, 1, 6, 198
Trade, 196, 209
Tradition, 100, 108, 117, 118, 128, 129, 213
Transitional
  forms, 104, 187, 215, 223, 225
  periods, 210
Transmission, 116, 117, 127, 128
Trenton (New Jersey), 166, 167
Trilobites, 68
Twelfth International Radiocarbon Conference, 15
Tylor, Edward B., 62
Types, 23, 26, 33, 34, 44, 54, 56, 57, 58; *see also* Decoration types; Historical, types; Pottery types; Projectile point types
  characteristics of, 23, 26, 207, 212
  for chronological use, 29, 35, 183, 196, 219, 220, 221, 222, 224
  distributions
    spatial, 44, 46, 54, 215
    temporal, 29, 44, 54, 111, 131, 187, 215
  kinds of, 24, 224
Types (*cont.*)
  reality of, 51, 55, 56, 99, 219, 220
Typological creep, 46, 91

Uhle, Max, 63, 64, 111, 160, 162, 163, 165, 220
Unimodal frequency distributions, 29, 58, 63, 116, 123, 124, 126, 129, 130, 163, 166, 167, 170, 171, 196, 220, 224
Units
  reality of, 21, 22, 51
  shifts in scale, 110
  used to measure change, 115
University of California, 26, 158
US National Museum, 2, 32, 158

Vaillant, George, 191, 192, 193, 196, 197, 198, 199, 212, 213
Valley of Mexico, 157, 158, 174, 192, 193, 197, 199, 222
Valley of the Kings, 84
Variation, 14, 28, 32, 41, 48, 55, 96, 99, 104, 117, 118, 126, 166, 187, 199, 226
Varve dating, 6
Vertebrata, 76
Vertically distinct units, 147
Viking Fund Medal, 63

Walls Engraved/Incised, 36
Walker, S. T., 153, 154
Wedel, Waldo, 221
Wheeler, Sir Mortimer, 140, 152
Willey, Gordon, 93, 100, 101, 149, 150, 179, 209, 213, 214, 223, 225
Willey and Phillips synthesis of prehistory, 224
Wilson, Thomas, 2
Wissler, Clark, 26, 161, 163, 165, 167, 172, 174, 192, 220, 221, 222
Worsaae, J. J. A., 11, 152
Wyman, Jeffries, 152, 153, 154

Yankee Blade component, 180
Yazoo River, 205
Yucatan, 192

Zacatenco (Mexico), 192, 193, 196
Zuñi Pueblo (New Mexico), 111, 112, 220